HEATING SERVICE

James L. Dundas
Professor of Applied Science

Macomb County Community College

KENDALL/HUNT PUBLISHING COMPANY
2460 Kerper Boulevard,
Dubuque, Iowa 52001

Illustrations by David Hoover

Library of Congress Catalog Card Number: 78—53665

ISBN 0—8403—1866—9

Printed in the United States of America

401866 01

Preface

This text on heating is different from most others, in that the author looks at the subject as a service man. There are many books on heating used in schools through out the country, but I have found that they are used primarily in school and seldom, if ever used again. So I have written this as if it were a service man's notebook, that he can use in the field, as well as the classroom. The text emphasizes on the equipment that is used in the field today, from equipment that is slightly outdated to the latest solid state controls. How to test, wire, install, service and trouble shoot heating equipment and accessories. The author's intent is that this book be used in colleges, technical schools, adult education classes and senior high schools. The text has been carefully written so that the reader will have little difficulty in comprehension.

James L. Dundas

Acknowledgments

The author acknowledges with deep gratitude and appreciation the cooperation of the **American Gas Association, Electro Air Division-Emerson Electric, Honeywell Inc., Bacharach Instrument Company, Dwyer Instruments Inc.**, and the **Auto Flo Company**. I feel that these companies are not just interested in the fine produce that they manufacture, but have a true interest in the student and in the future serviceman.

Contents

Chapter 1

HIGH PRESSURE ATOMIZING OIL BURNERS

High-pressure atomizing burners consist of

1. Fuel system, including pump, pressure-regulating valve and cut-off fuel nozzle, strainers, piping
2. Air system, including fan, air tube, air head, vanes, air shutter
3. Ignition system, including transformer, electrodes and wiring
4. Motor

Rotary gear pump is a positive displacement pump. Fuel is pushed through it by gear teeth and must have a place to go or serious damage will result.

Rotation must be correct; reversed rotation will give no oil delivery. Most units have correct rotation indicated by an arrow on the pump casting.

The *gross capacity* of an oil pump is always in excess of the firing rate—usually many times in excess.

Pressure is maintained to a desired point by the *pressure-regulating* valve, which is a spring-loaded valve opening with the increase of pressure, thus holding pressure. A reverse action occurs when the pressure falls, the valve will close and stop the oil from going to the nozzle.

Pressure-regulating valve may consist of plunger that uncovers a by-pass port as it is forced down its cylinder against spring pressure; or of a needle valve seating in an orifice in the face of a bellows or plunger. Adjustment is by changing spring compression. Usually design includes a cover or cap, which, if removed, gives access to a slotted pressure-regulating screw. This can be turned by a screwdriver to increase or decrease pressure.

Cut-off valve is usually combined with pressure-regulating valve. It is a spring-loaded valve opening with a rise in pressure—usually set to open at 70 to 80 lbs. When open, it allows fuel to flow to nozzle.

Pressure should be adjusted only when the nozzle used is connected by the regular oil line, and the pump is up to speed. A pressure gauge should be used, as it is impossible to guess pressure.

Some pressure-regulating valves have a slot cut in the needle valve, which makes an orifice always open, yet too small to handle the by-passed capacity without opening the valve. This orifice helps to vent any accumulated air.

The by-pass of fuel through a pressure-regulating valve may be through a port inside the pump back to the suction side; or through external return piping back to the fuel tank. The first is a *one-pipe system;* second is a *two-pipe system.*

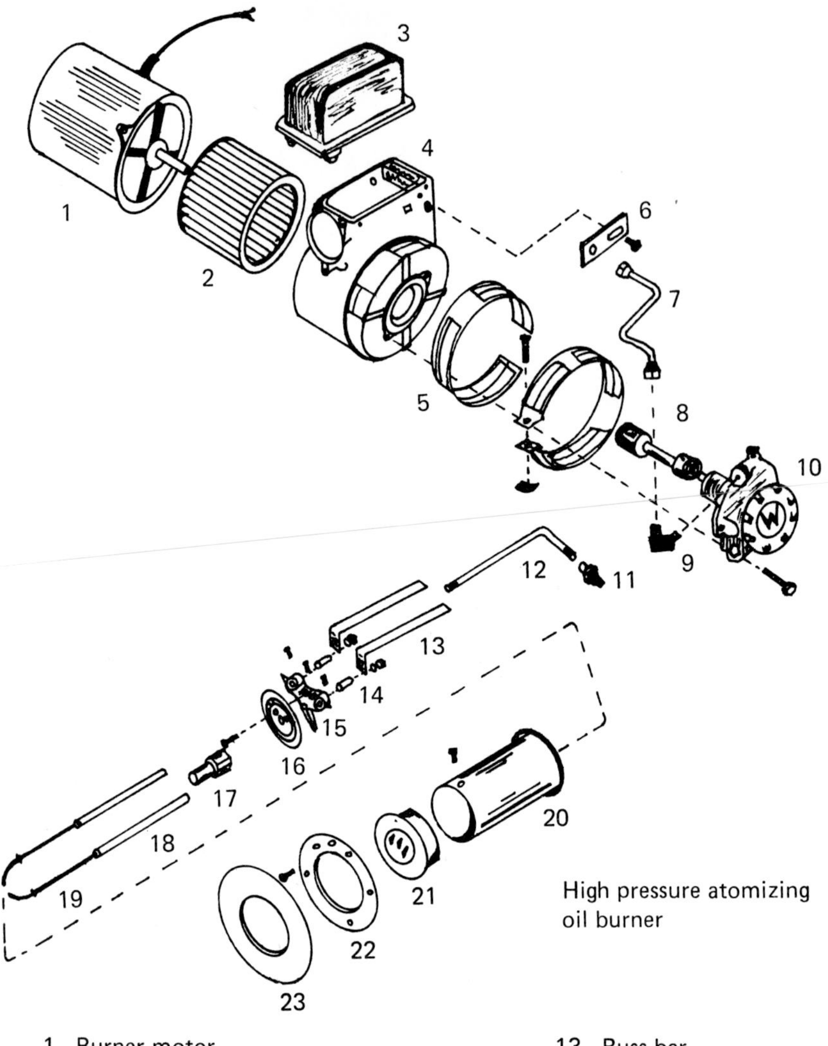

1. Burner motor
2. Fan
3. 10,000 volt transformer
4. Burner housing
5. Air adj. band
6. Oil line slot cover
7. Oil line outlet
8. Pump coupling
9. Oil line elbow
10. Fuel unit
11. Oil line fitting
12. Oil pipe
13. Buss bar
14. Insulator bushing
15. Electrode support
16. Baffle plate
17. Nozzle adapter
18. Insulators
19. Electrodes
20. Air tube
21. Air cone
22. Flange
23. Gasket

Figure 1.1. High pressure atomizing oil burner.

Fuel units have a tapped opening for an external by-pass line, if closed with a pipe plug it is arranged for a one-pipe system. In change-over to two-pipe system you must have a 1/8" by-pass plug inserted in the internal port, and connect the return line into the tapped opening. Then from the tapped opening on the pump to the top of the oil tank, the return line must then go within a few inches of the bottom.

It is vitally important that the internal port be free for a one-pipe system. It is equally important to have the internal port plugged and the return line connected for a two-pipe system.

One-pipe systems are permissible if the fuel tank is located in basement or above grade or will flow to the pump by gravity. A two-pipe system is vital if the tank is buried, or inside but below pump level.

Fuel units may be of a single-stage or a two-stage design. With the two-stage, there are two pumps, one picks up oil from the tank and discharges it to an intermediate chamber, at or about atmospheric pressure; the second pump draws from the intermediate chamber and discharges it to the nozzle. By dividing the total pressure between two pumps, a two-stage unit is capable of higher suction lifts.

The suction lift is limited by atmospheric pressure and tendency of oil to vaporize at higher vacuums. Although pumps differ, generally single-stage pumps should not be installed for over 12 feet static lift, although two-stage units are sometimes installed with as much as 20 feet static lift. These units are based on reasonable length and size of suction line. For units used in domestic installations, a reasonable length is 25 feet and pipe size should not be smaller than 3/8-in. OD tubing or 3/8-in. iron-pipe size.

Total suction lift may be checked with a vacuum gauge, connected to the suction side of the unit. Usually there is a tapped opening with a pipe plug. When the pump is operating normally, the vacuum indicates total suction lift, or static lift plus friction. One inch of vacuum is approximately equal to 1-1/3 ft. of 30 gravity fuel.

The suction line must be tight. If it is not, air will be drawn into fuel oil and there will be a fluctuating flame, poor cut-off or flame failure. Any of these symptoms can be checked against leaking suction lines by operating the burner with a temporary line drawing from a pail of fuel oil. If the symptoms stop, the indication is that leaks may be responsible.

A suction line must be unobstructed. Care should be taken to see that copper tubing is not kinked when installing and that anti-syphon valves or check valves are freely operating. Hand-operated valves should be fully open.

Restrictions in a suction line may cause vapor binding, resulting in the same symptoms as for air leaks. A temporary line will, therefore, confirm your check out on air leaks or obstructions but will not distinguish between them.

A vacuum gauge reading, however, will indicate obstructions. If the static lift is, for example, 5 feet, but the running vacuum reads 15 in., the indication is that friction lost is 13 ft.—entirely too much. On the other hand, if the trouble is leakage, the vacuum gauge reading would reflect a normal friction loss as long as there is any oil delivery.

Rotary gear pumps rely on the fuel oil being handled for *internal lubrication* and *sealing.* A dry pump may pull so little vacuum that oil cannot be picked up with only a slight static lift. Continued operation in the hopes of starting may injure the pump. It is wise to *prime* any dry fuel unit before starting up with a suction lift. This can usually be done by removing a cover or plug on the pump compartment and filling the unit with fuel. A pump supplied with fuel by gravity presents no such problem—but hand-operated valves in the line should be checked to see that they are open. Most units that are started up dry, or that have been opened for any reason, should be *air vented* on starting. Two-stage units with return line to the tank may be an exception. Venting is done by loosening or removing the pressure-gauge plug and operating the pump until about a pint of liquid oil is discharged.

It is essential that a pump on a suction lift *retains its prime* during off periods. One way is to have a *foot valve* on the end of the suction line in the tank. Some engineers and servicemen do not like foot valves, as they are difficult to inspect them. They prefer an easily accessible *check valve*—which usually means a check valve at the top of the tank or in the line after it enters the basement.

It is difficult to prevent all leakage in either foot valves or check valves, but there are several means for minimizing leakage. One is to extend the return line almost to the bottom of the tank (but not close to the suction line, as otherwise sediment might be stirred up). This seals the return, preventing air from working it up to the pump, and constitutes a barometric loop.

A strainer occasionally needs cleaning. Fuel units have covers that provide access and opportunity for removal. Strainers may be cleaned in a solvent. The compartment should be cleaned internally at the same time. A strainer has a perforated metal screen or wire mesh for an element. They are rated in number of openings per square inch—"100 mesh," "120 mesh," etc. For domestic burners, suction strainer is usually of 100 mesh.

Filters are sometimes used in fuel units instead of strainers, and are sometimes installed externally in conjunction with fuel-unit strainers. A filter has a replaceable element. Passage of fuel through a filter is considerably more difficult than it is through a strainer. A filter should have considerably greater filtering surface. Practically all filters have replaceable elements. The element should be replaced rather than cleaned.

Several devices are used to insure sharp cut-off when the burner stops, as after-drip may carbonize the nozzle, cause flame pulsation at shutdown, or a blow-back on restarting. One such device is an electrically-controlled solenoid valve. The electromagnet is wired in parallel with the burner motor. When energized, it takes time to open, or is delayed, this permits pressure to build up.

When de-energized, as the motor is cut off, it will close and cause pressure to drop suddenly, thus insuring a good clean cut off of the oil to the nozzle.

Pressure-regulating and cut-off assemblies may need *cleaning*. Need for cleaning may be indicated by a fluctuating flame, abnormally large flame, small sparking flame, retarded cut-off or after-burn. Usually a cover plate gives access to the pressure-regulating assembly and removing of oil line and screw plug in which it is screwed frees the cut-off assembly. Parts may be cleaned in a solvent. If the plunger of a plunger-type valve is worn, it should be replaced, and it may be necessary to replace the sleeve also. If the bellows of a bellows-type valve is defective in any way, the bellows assembly should be replaced. Flex a bellows close to the ear; if broken or cracked, the flexing bellows will give a clicking noise.

The fuel unit delivers fuel oil at or about 100 lbs. through the oil pipe to the adapter and the *nozzle*. One function of the nozzle is to act as a high-pressure *metering orifice*, as the amount of oil passed by the nozzle will depend on the nozzle design, size of passages, fuel oil pressure and viscosity.

Typical nozzle consists of *strainer, plug with slots, shell with orifice*. Fuel must first pass through the strainer. Averages are 100 mesh for 1.35 gph and larger, 120 mesh for 1.00 to

1.35 gph and 200 mesh for less than 1.00 gph. Fuel from the strainer passes through or around the plug and down the slots to the orifice.

The slots are not radial but tangential, to give the fuel a rapid rotation in the whirl chamber and through the orifice. The effect of the rotary motion alone would be to whirl the fuel off the outer edge of the orifice in the shape of a flat disc. The fuel has forward velocity also, however, so the disc is modified into a hollow cone. The angle of the cone will be a function of the amount of whirl and the forward velocity.

It is practical to use firing pressures in the 95 to 105 lb range, but the range of 90 to 110 lb. is a pressure that may be unsatisfactory. The lower pressure may not atomize satisfactorily and cut-off action may be erratic; the higher one may put too great a load on the pump.

Higher pressures result in finer atomization but less spray travel; lower pressures result in a coarser spray—harder to vaporize—but the drops travel further. Viscosity has an effect of capacity, spray angle and atomization.

Changing the nozzle position axially forward in the air tube widens the flame, but too much forward movement will result in nozzle overheating and carbonization. Drawing the nozzle back will lengthen the flame, but if withdrawn too far, the nozzle will spray in the air tube.

Although firing rate is stamped on a nozzle, this based on a 35 SSU fuel oil at 100 lb. pressure, actual delivery rate may exceed or fall short of the stamped rate by five percent, especially for nozzles of low firing rates. Such nozzles have orifices of 0.010-in. diameter of less and slots only 0.0005 in. deep. A microscopic variation in such dimensions will result in considerable performance variation. It is for this reason that a serviceman should carry an assortment of nozzles when he goes to adapt a burner to a combustion chamber.

Nozzles are also calibrated for angle of spray, but exact angle of actual fire cone is result of combined action of nozzle and air flow. Consequently, an 80 degree nozzle may result in a fire cone larger or smaller than 80 degrees in an actual burner firing in actual combustion chamber.

Nozzles are made to watch-like precision and should be handled carefully and with appropriate tools. Nozzles should be replaced and not cleaned. Nozzles should be left in the shipping cartons until applied. Clean tools should be used in installing.

The ignition system consists of an *ignition transformer,* porcelain *insulators, electrodes,* and *wiring.* The usual type of transformer increases a voltage from 110v to 10,000v. They are special transformers, known as "high reactance" type. With a transformer not of this type, the current would build up indefinitely and cause damage to the electrical system.

Shields are also placed between primary and secondary windings to eliminate transfer of radio frequency interference. Such shields are grounded to the case, and the case is grounded externally. The secondary winding may also be grounded at the mid point. After manufacture, the core and windings are sealed by filling the case with an insulating plastic that is solid when at ordinary temperature.

Transformers that have become water-soaked should be thoroughly dried before current is put through them. If they are still damp, normal voltage may burn them out.

The *spark gap* provided by the electrodes is normally *1/8 in. wide.* It is usually centered about *1/2 in. above the nozzle orifice and about 1/8 in. ahead of it.* The aim is to keep the electrode tips about 1/8 in. out of the fuel spray, relying on the air velocity to extend the actual spark into the fuel.

Electrodes that are closer to metal parts of the air tube than by a distance equal to two times the spark gap, or 1/4 in. for 1/8 in. gap, may develop accidental grounds. Electrodes are run through porcelain insulators, which in turn are supported by clamps. Minute cracks in such insulators sometimes cause grounding. Some such cracks will ground the current only at higher temperatures.

The Air System is made up of

- centrifugal fan
- fan housing
- intake shutter or band
- air tube (gun tube)
- vanes and air head

Combustion air is furnished by a centrifugal fan of 5 to 6 in. diameter, usually with forwarding-curving blades to increase air velocity. The fan is enclosed in housing and suction air is introduced by the way of an adjustable opening in the form of a movable band or shutter. The fan should supply sufficient air for good combustion when the entrance openings are not fully open.

The air is delivered down the air tube and around the nozzle and is usually given a turbulent flow by spiral vanes on the inner surface of the air tube.

TEST YOUR KNOWLEDGE

1. A Rotary pump is a positive displacement pump. (T) or (F).
2. The gross capacity of an oil pump is always in excess of the firing rate. (T) or (F).
3. The pressure is maintained to a desired point by the ____________________ .
4. The cut-off valve is usually combined with the pressure-regulating valve. (T) or (F).
5. Some pressure-regulating valves have a slot cut in the needle valve, which makes an orifice always open. (T) or (F).
6. It is vitally important that the internal port be closed. (T) or (F).
7. Generally single-stage pumps should not be installed over ________________ of static lift.
8. One inch of vacuum is approximately equal to ________________ ft. of 30 gravity fuel oil.
9. The fuel units oil strainer occasionally needs cleaning. (T) or (F).
10. A strainer has a metal screen or wire mesh. They are rated in the number or openings per square inch. The rating would be ____________________ .
11. When the burner motor is de-energized the fuel unit will open and cause a pressure drop. (T) or (F).
12. The fuel unit delivers fuel at or about ____________________ .
13. Rotary pumps rely on the fuel oil being handled for internal lubrication. (T) or (F).
14. Suction lines must be tight. (T) or (F).
15. Restrictions in a suction line may cause ____________________ .

NOZZLES

How the nozzle fits into the performance of an oil burner. Like most combustible material, the oil must first be vaporized or converted to a vapor or gas before combustion can take place. This is usually accomplished by the application of heat. The oil vapor must be mixed with air in order to have oxygen present for combustion. The temperature of this mixture must be increased above the ignition point. A continuous supply of air and fuel must be provided for continuous combustion and the products of combustion must be removed from the combustion chamber. The atomizing oil nozzle performs three basic functions.

1. *Atomizing*
   Atomizing speeds up the vaporization process by breaking up the oil into tiny droplets. The exposed surface of a gallon of oil is thereby expanded to approximately 690,000 square inches of burning surface. Individual droplet sizes range from .0002 inch to .010 inch. The smaller droplets are necessary for fast, quiet ignition and to establish a flame front close to the burner head. The larger droplets take longer to burn.
2. *Metering*
   The nozzle is designed and sized, so that it will deliver a fixed amount of atomized fuel to the combustion chamber, within a plus or minus range, of 5% of the rated capacity.
3. *Patterning*
   The nozzle is also expected to deliver the atomized fuel to the combustion chamber in a uniform spray pattern and spray angle best suited to the requirements of a specific burner.

The proper nozzle selection is of great importance, because the performance of the nozzle is so directly related to the over-all performance of the burner. The wrong choice of flow rate, spray angle or spray pattern for a given burner's air pattern can result in improper firing. See Figure 1.6.

Hollow Cone Pattern

The hollow cone nozzle as the name implies, has the greatest concentration of droplets at the outer edge of the spray, with little or no droplets distribution in the center. In general, they can be recommended for use in the smaller oil burners. Those with firing rates of 1.00 GPH and

Nozzle spray patterns

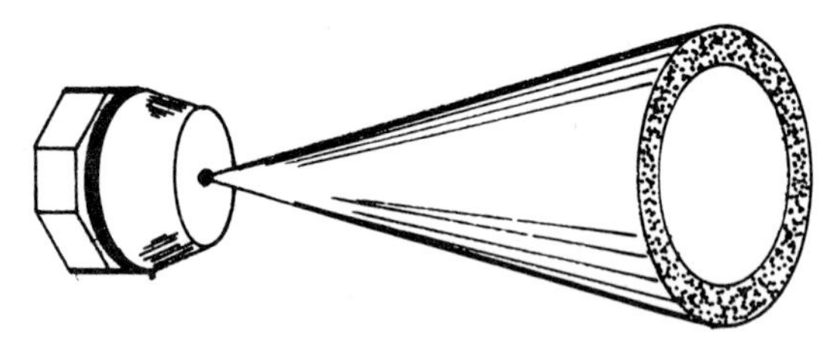

Figure 1.2. Hollow cone.

under, hollow cone nozzles, generally have more stable spray angles and patterns under adverse conditions than solid cone nozzles of the same flow rate. See Figure 1.2.

Solid Cone Pattern

The Solid Cone Nozzle (With the solid cone)

The distribution of droplets is fairly uniform throughout the pattern. This pattern is recommended where the air pattern of the oil burner is heavy in the center, or where long fires are required, or for smoother ignition in most burners firing above 2.00 GPH. See Figure 1.3.

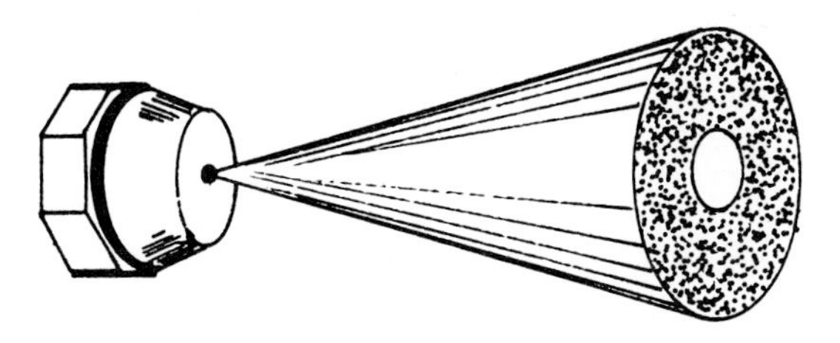

Figure 1.3. Solid cone.

All-purpose Pattern

The All-purpose Nozzle

An all-purpose nozzle is neither a true hollow cone nor a true solid cone nozzle. At the lower flow rates, it tends to be more hollow than solid, and as the flow rate increases, the pattern becomes more like the solid cone. It can be used in place of solid or hollow cone nozzles between .50 GPN and 8.00 GPH. See Figure 1.4.

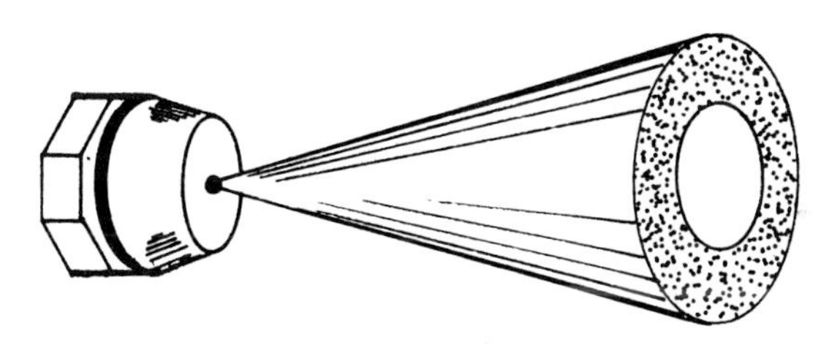

Figure 1.4. All-purpose.

Variety of Spray Angles

This is the angle of the cone of spray. Spray angles are generally available from 30 degrees to 90 degrees to meet the requirements of a wide variety of burner air patterns and combustion chambers. Some burners can only use one spray angle for good efficiency. For the sizing of nozzles, see Figure 1.5.

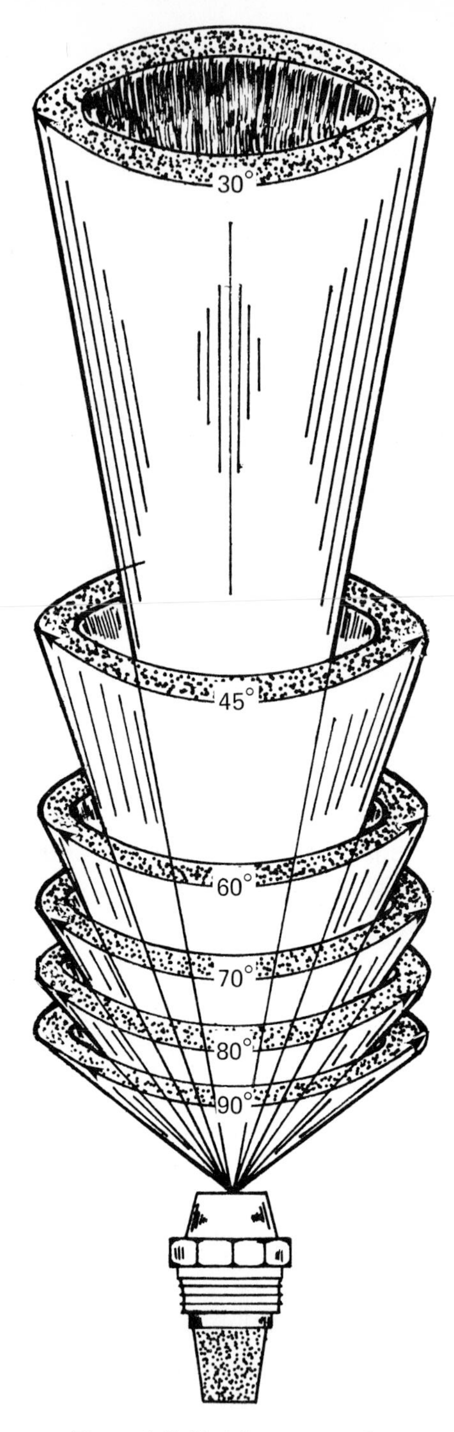

Figure 1.5. Nozzle spray angle.

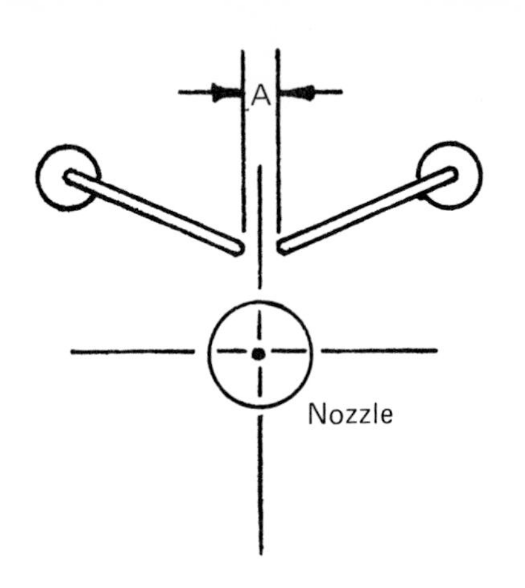

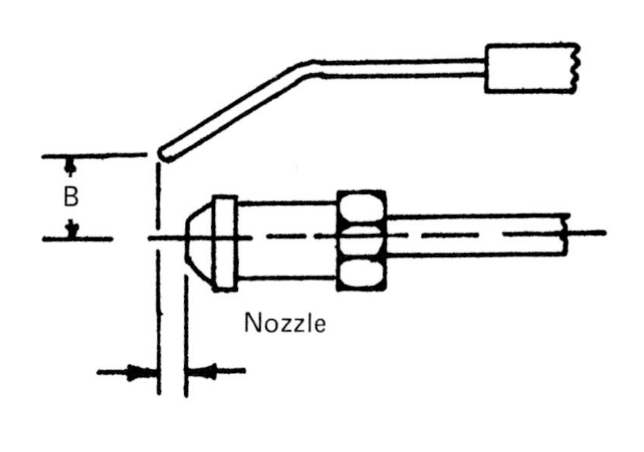

Nozzle	G.P.H.	A	B	C
45°	.75 to 3.00	1/8″ to 3/16″	1/2″ to 9/16″	1/4″
60°	.75 to 3.00	1/8″ to 3/16″	9/16″ to 5/8″	1/4″
70°	.75 to 3.00	1/8″ to 3/16″	9/16″ to 5/8″	1/8″
80°	.75 to 3.00	1/8″ to 3/16″	9/16″ to 5/8″	1/8″
90°	.75 to 3.00	1/8″ to 3/16″	9/16″ to 5/8″	0

Figure 1.6.

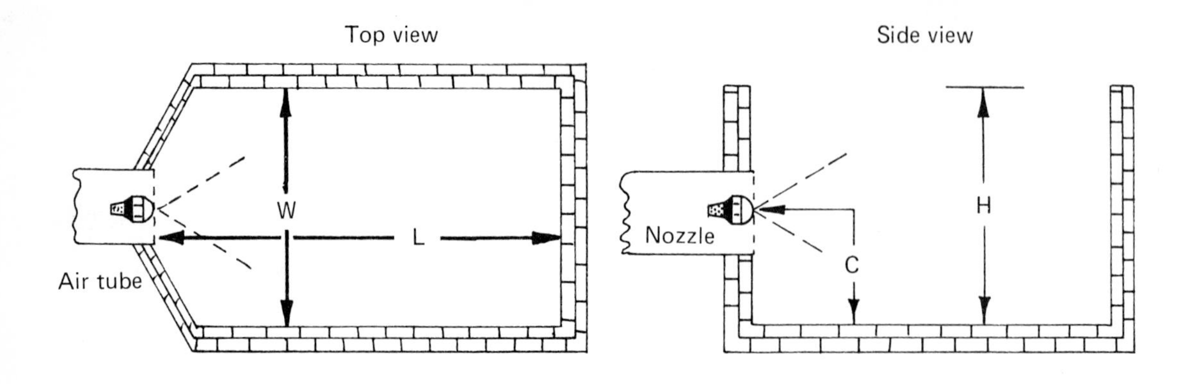

Nozzle Size or Rating (GPH)	Spray Angle	Square or Rectangular Combustion Chamber				Round Chamber (diameter in inches)
		L Length (in)	W Width (in)	H Height (in)	C Nozzle Height (in)	
.50- .65	80°	8	8	11	4	9
.75- .85	60°	10	8	12	4	*
	80°	9	9	13	5	10
1.00-1.10	45°	14	7	12	4	*
	60°	11	9	13	5	*
	80°	10	10	14	6	11
1.25-1.35	45°	15	8	11	5	*
	60°	12	10	14	6	*
	80°	11	11	15	7	12
1.50-1.65	45°	16	10	12	6	*
	60°	13	11	14	7	*
	80°	12	12	15	7	13
1.75-2.00	45°	18	11	14	6	*
	60°	15	12	15	7	*
	80°	14	13	16	8	15
2.25-2.50	45°	18	12	14	7	*
	60°	17	13	15	8	*
	80°	15	14	16	8	16
3.00	45°	20	13	15	7	*
	60°	19	14	17	8	*
	80°	18	16	18	9	17

Figure 1.7.

Nozzle Rating at 100 PSI	Nozzle Flow Rates in Gallons per Hour						Nozzle Rating at 100 PSI	Nozzle Flow Rates in Gallons per Hour					
	80 PSI	120 PSI	140 PSI	160 PSI	200 PSI	300 PSI		80 PSI	120 PSI	140 PSI	160 PSI	200 PSI	300 PSI
.50	.45	.55	.59	.63	.70	.86	7.50	6.65	8.20	8.85	9.50	10.6	13.0
.65	.58	.71	.77	.82	.92	1.12	8.00	7.10	8.75	9.43	10.1	11.3	13.8
.75	.67	.82	.89	.95	1.05	1.30	8.50	7.55	9.30	10.0	10.7	12.0	14.7
.85	.76	.93	1.00	1.08	1.20	1.47	9.00	8.00	9.85	10.6	11.4	12.7	15.6
.90	.81	.99	1.07	1.14	1.27	1.56	9.50	8.45	10.4	11.2	12.0	13.4	16.4
1.00	.89	1.10	1.18	1.27	1.41	1.73	10.00	8.90	10.9	11.8	12.6	14.1	17.3
1.10	.99	1.21	1.30	1.39	1.55	1.90	11.00	9.80	12.0	13.0	13.9	15.5	19.0
1.20	1.07	1.31	1.41	1.51	1.70	2.08	12.00	10.7	13.1	14.1	15.1	17.0	20.8
1.25	1.12	1.37	1.48	1.58	1.76	2.16	13.00	11.6	14.2	15.3	16.4	18.4	22.5
1.35	1.21	1.48	1.60	1.71	1.91	2.34	14.00	12.4	15.3	16.5	17.7	19.8	24.2
1.50	1.34	1.64	1.78	1.90	2.12	2.60	15.00	13.3	16.4	17.7	19.0	21.2	26.0
1.65	1.48	1.81	1.95	2.09	2.33	2.86	16.00	14.2	17.5	18.9	20.2	22.6	27.7
1.75	1.57	1.92	2.07	2.22	2.48	3.03	17.00	15.1	18.6	20.0	21.5	24.0	29.4
2.00	1.79	2.19	2.37	2.53	2.82	3.48	18.00	16.0	19.7	21.2	22.8	25.4	31.2
2.25	2.01	2.47	2.66	2.85	3.18	3.90	19.00	16.9	20.8	22.4	24.0	26.8	33.0
2.50	2.24	2.74	2.96	3.16	3.54	4.33	20.00	17.8	21.9	23.6	25.3	28.3	34.6
2.75	2.44	3.00	3.24	3.48	3.90	4.75	22.00	19.6	24.0	26.0	27.8	31.0	38.0
3.00	2.69	3.29	3.55	3.80	4.25	5.20	24.00	21.4	26.2	28.3	30.3	34.0	41.5
3.25	2.90	3.56	3.83	4.10	4.60	5.63	26.00	23.2	28.4	30.6	32.8	36.8	45.0
3.50	3.10	3.82	4.13	4.42	4.95	6.06	28.00	25.0	30.6	33.0	35.4	39.6	48.5
4.00	3.55	4.37	4.70	5.05	5.65	6.92	30.00	26.7	32.8	35.4	38.0	42.4	52.0
4.50	4.00	4.92	5.30	5.70	6.35	7.80	32.00	28.4	35.0	37.8	40.5	45.2	55.5
5.00	4.45	5.46	5.90	6.30	7.05	8.65	35.00	31.2	38.2	41.3	44.0	49.5	60.5
5.50	4.90	6.00	6.50	6.95	7.78	9.52	40.00	35.6	43.8	47.0	50.5	56.5	69.0
6.00	5.35	6.56	7.10	7.60	8.50	10.4	45.00	40.0	49.0	53.0	57.0	63.5	78.0
6.50	5.80	7.10	7.65	8.20	9.20	11.2	50.00	44.5	54.5	59.0	63.0	70.5	86.5
7.00	6.22	7.65	8.25	8.85	9.90	12.1	55.00	48.0	62.0	66.0	68.0	73.0	89.0

Figure 1.8. Effects of pressure on nozzle flow rate.

Nozzle Care and Handling

Keep nozzles in a clean place. In a suitable rack or box. Nozzles should not be permitted to roll around in a drawer. When carrying nozzles in a truck or tool box, keep them in a clean steel box provided by the nozzle manufacturer for that purpose. Handle the nozzle carefully after removing it from its individual vial. Don't touch the screen with dirty hands because grease or dirt can be squeezed through the screen, thus contaminating the nozzle. Be sure the strainer or filter is in place on the nozzle but do not take the nozzle apart or remove the strainer before installing it. The nozzle manufacturer took great care to be sure the nozzle is clean. The nozzle orifice is finished to a glass-like finish by the manufacturer. Don't ruin it with a wire or pin or by bumping it with a wrench. Be sure the burner tube is flush with the combustion chamber wall and is well insulated. If the burner extends into the chamber, the nozzle will become overheated and will carbon up in a short time. Never try to clean a sludged up nozzle, it is not profitable because of the time it takes and it frequently results in no-charge call-backs. The contamination loosened in cleaning works its way into the feed slots and orifice of the nozzle. You should replace nozzles annually. After a nozzle has been fired for a full heating season, it is a sound service practice to change it even though it is working O.K. at the time of your service call.

Nozzle	Total Load		
Nozzle Size GPH	Hot Water Radiation (feet)	Steam Radiation (feet)	Warm Air (BTU/hr. at bonnet)
.75	446	282	67,100
.85	506	319	76,060
1.00	595	375	89,500
1.20	715	450	107,500
1.35	805	505	120,900
1.50	895	560	134,400
1.65	985	615	147,800
2.00	1,195	745	179,200
2.50	1,495	935	224,000
3.00	1,790	1,120	268,800
3.50	2,090	1,305	313,600
4.00	2,390	1,490	358,400
5.00	2,985	1,865	448,000
6.00	3,580	2,240	537,600
7.00	4,180	2,680	627,200
8.00	4,775	2,985	716,800
9.50	5,670	3,545	851,200
10.50	6,270	3,920	940,800

Figure 1.9.

TEST YOUR KNOWLEDGE

1. Atomizing speeds up the vaporization process by breaking up the oil into large droplets. (T) or (F).
2. The nozzle is designed to deliver a fixed amount of fuel, within a plus or minus range of

   ______________ .
3. The three basic functions that the nozzle performs. 1. ______________________ ,

   2. ____________________________ , 3. ____________________________________ .
4. The all-purpose nozzle is the same as the solid cone nozzle. (T) or (F).
5. Spray angles are generally available from __________________ to __________________ .
6. Keep nozzles clean. (T) or (F).
7. You can clean a nozzle with a wire. (T) or (F).
8. Some burners can only use one spray angle. (T) or (F).
9. Nozzle selection is of great importance. (T) or (F).
10. Individual oil droplet sizes range from .02 to .1 inch. (T) or (F).

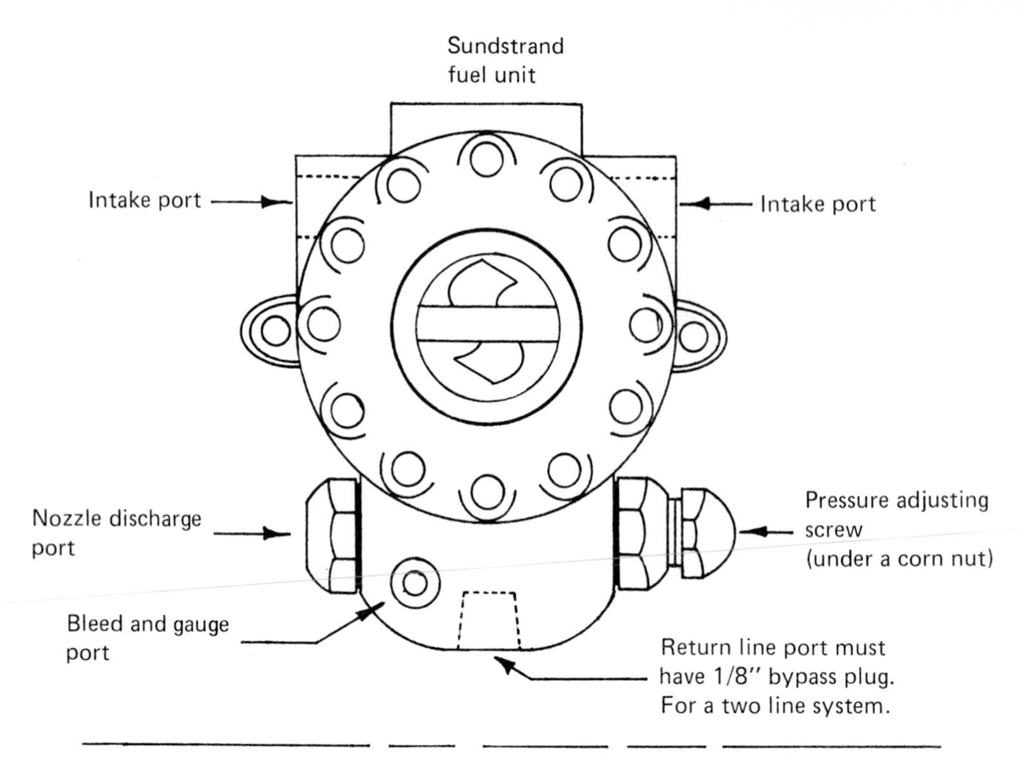
Sundstrand
fuel unit
Intake port
Intake port
Nozzle discharge
port
Pressure adjusting
screw
(under a corn nut)
Bleed and gauge
port
Return line port must
have 1/8″ bypass plug.
For a two line system.

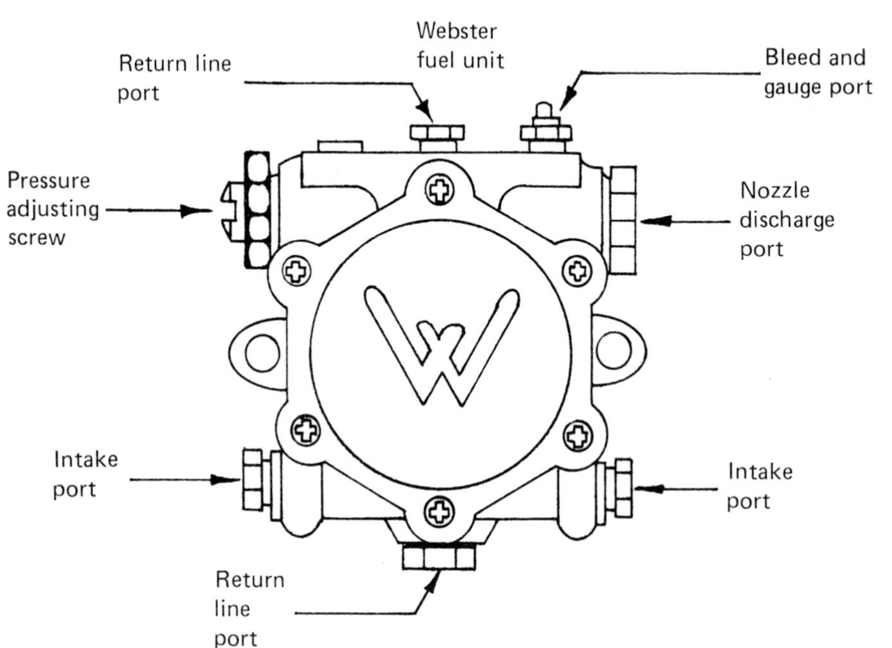
Webster
fuel unit
Return line
port
Bleed and
gauge port
Pressure
adjusting
screw
Nozzle
discharge
port
Intake
port
Intake
port
Return
line
port

Figure 1.10.

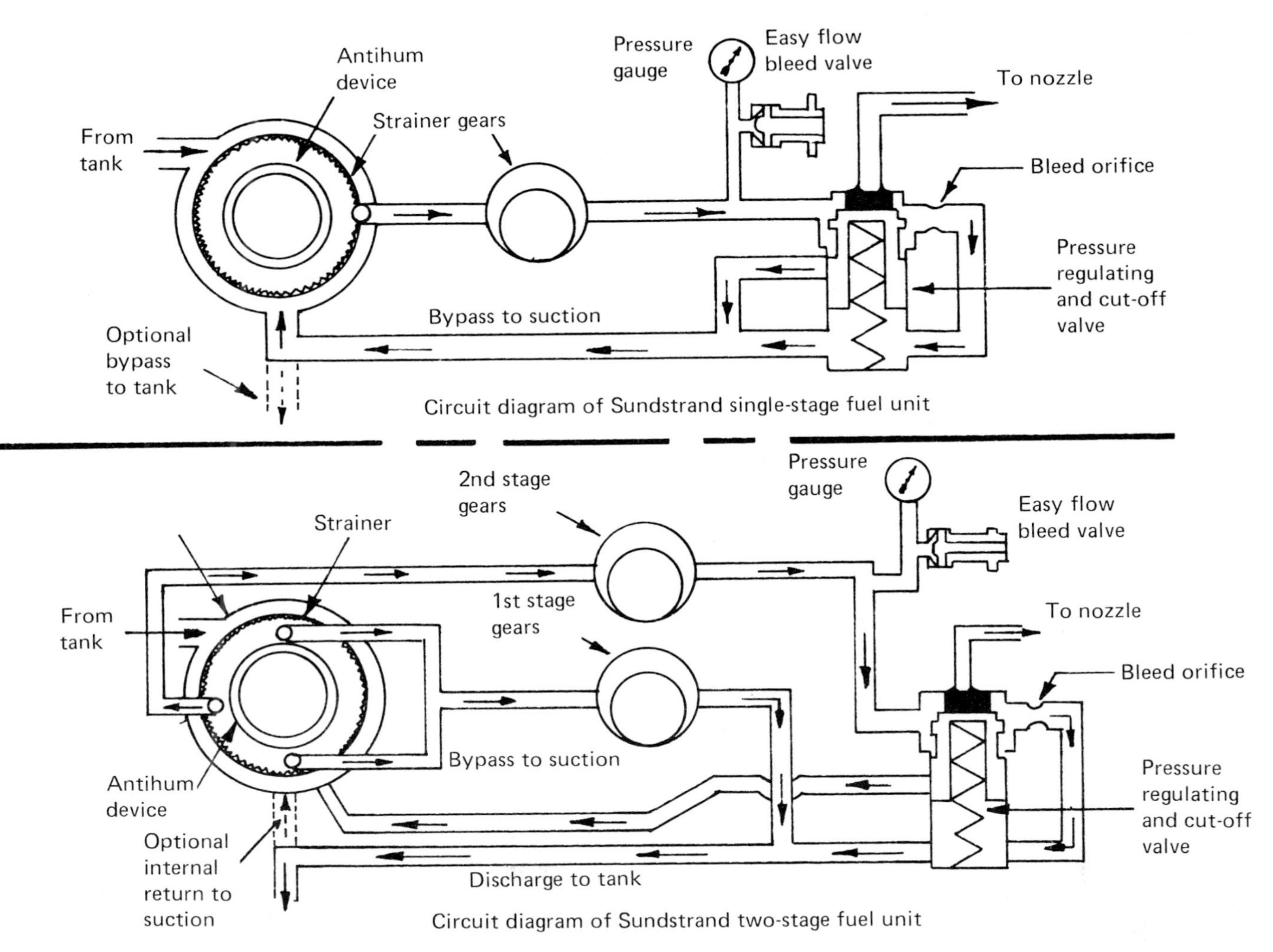

Circuit diagram of Sundstrand single-stage fuel unit

Circuit diagram of Sundstrand two-stage fuel unit

Figure 1.11.

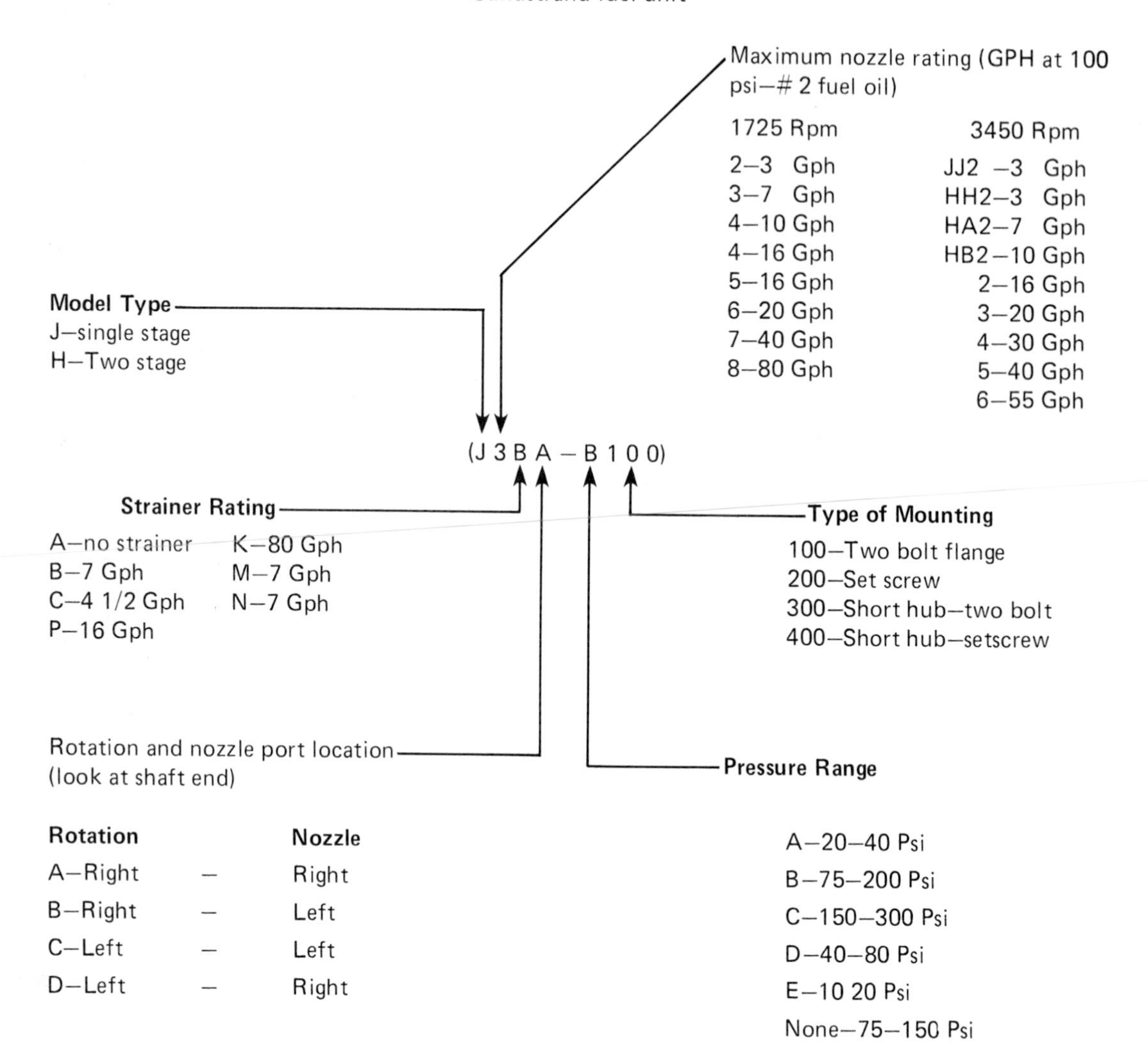

OIL TANKS AND PIPING

The oil storage tank may be located either inside or outside the building and either above or below the burner level. The approved construction of outside storage tanks differs from that for inside tanks. Usually the governing ordinance, municipal or state, specifies the steel gauge and maximum size and number of tanks which may be placed in the basement of a building and provides regulations pertaining to the construction, location and installation, either inside or outside. Most ordinances permit at least one and usually two inside storage tanks, 14 ga. 275 gals. capacity. However, all installations should conform to the governing ordinance.

For the average installation of an inside gravity tank of 275 gal., use a single stage fuel unit with a one pipe system, with 3/8" OD tubing from the tank to the burner. The tank should have four support legs of not less than 1 1/4″ ID. and not less than six inches in length. The tank should have a fusible valve on the outlet. The vent pipe should be connected to the top of the tank and vent *outside* at least two feet above grade and terminate at least two ft. from any door, window, overhang, or opening. The vent pipe must be at least 1 1/4″ black iron pipe. The fill pipe should connect to the top of the oil tank, and go out near the vent pipe. The fill pipe should be at least 18″ above grade, and be not less than 1 1/2″ black iron pipe. Gravity tanks should have a vent alarm installed between the tank and the vent pipe. Gravity tanks should have an oil level gauge installed in the top of the tank. For the installation of gravity tanks see Figure 1.12.

Inside Tank Installation Non-Gravity

For the average inside tank installation, when the lift from the tank to the burner does not exceed 10′, and the length of the suction line is not exceeding 30′, use a single stage or two stage fuel unit with a two pipe system and 3/8″ OD tubing. You must install a vertical ball check valve in the suctionline. If a single stage unit is used, even with a two pipe system, manual venting is required when the burner is first started, or if the tank runs dry. A two stage unit is self-venting with a two pipe system.

Installation of a Buried Tank

Use a two pipe system with either a single or two stage fuel unit, depending on the horizontal distance and lift. The maximum recommended lift for a single stage unit is 10 feet, and not over 30 feet of suction line (3/8″ tubing). Run the return line to 3″ from the bottom of the tank. For a lift installation of over 10 feet, use a vertical ball check valve in the suction line, inside the building. See Figure 1.13 for buried oil tanks.

Size of Tubing from the Tank to Burner

3/8″ OD tubing for vertical lifts up to 20 feet, and horizontal runs not exceeding 50 feet.
1/2″ OD tubing for dual burner installations, or runs exceeding 50 feet.

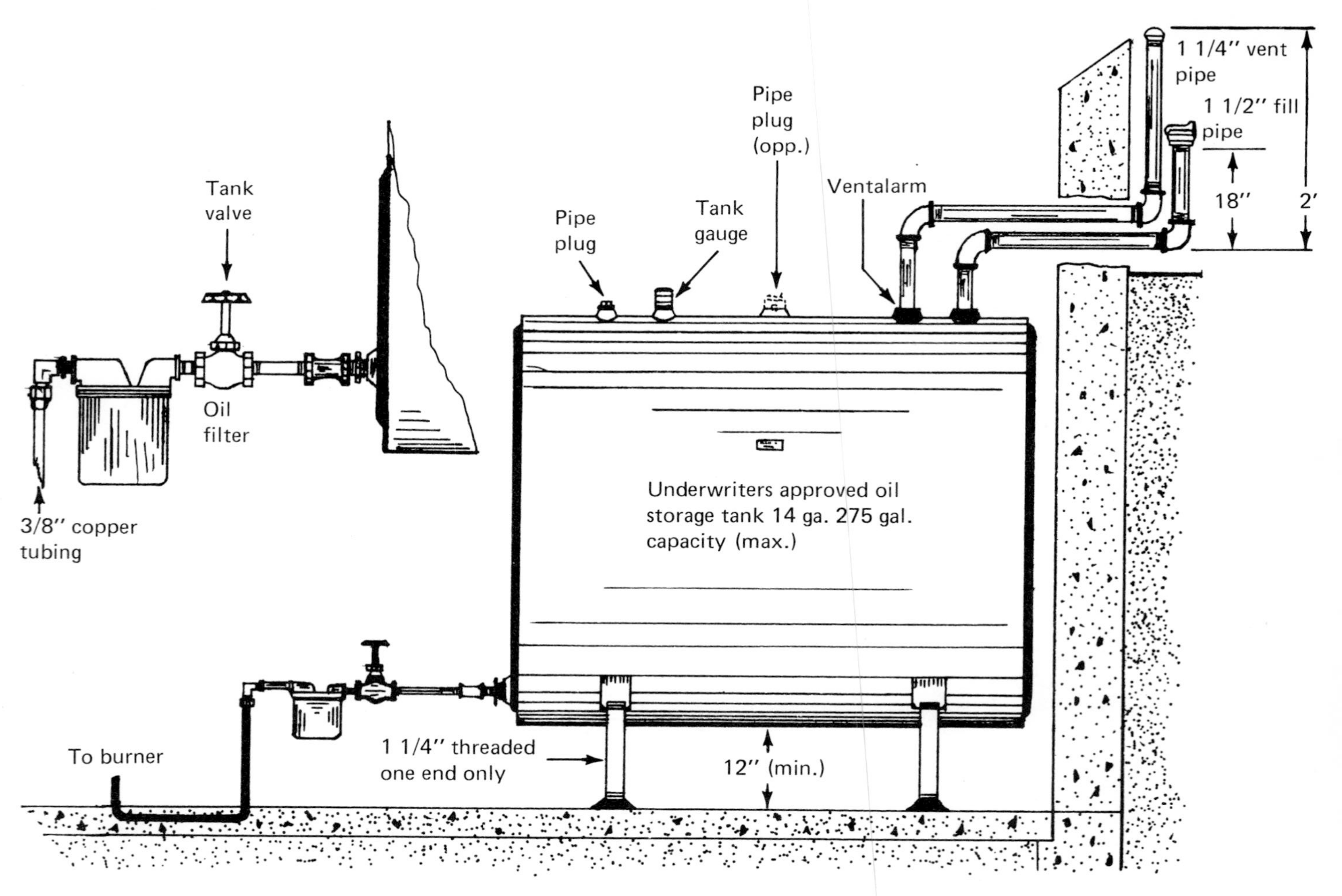

Figure 1.12. Single inside tank installation.

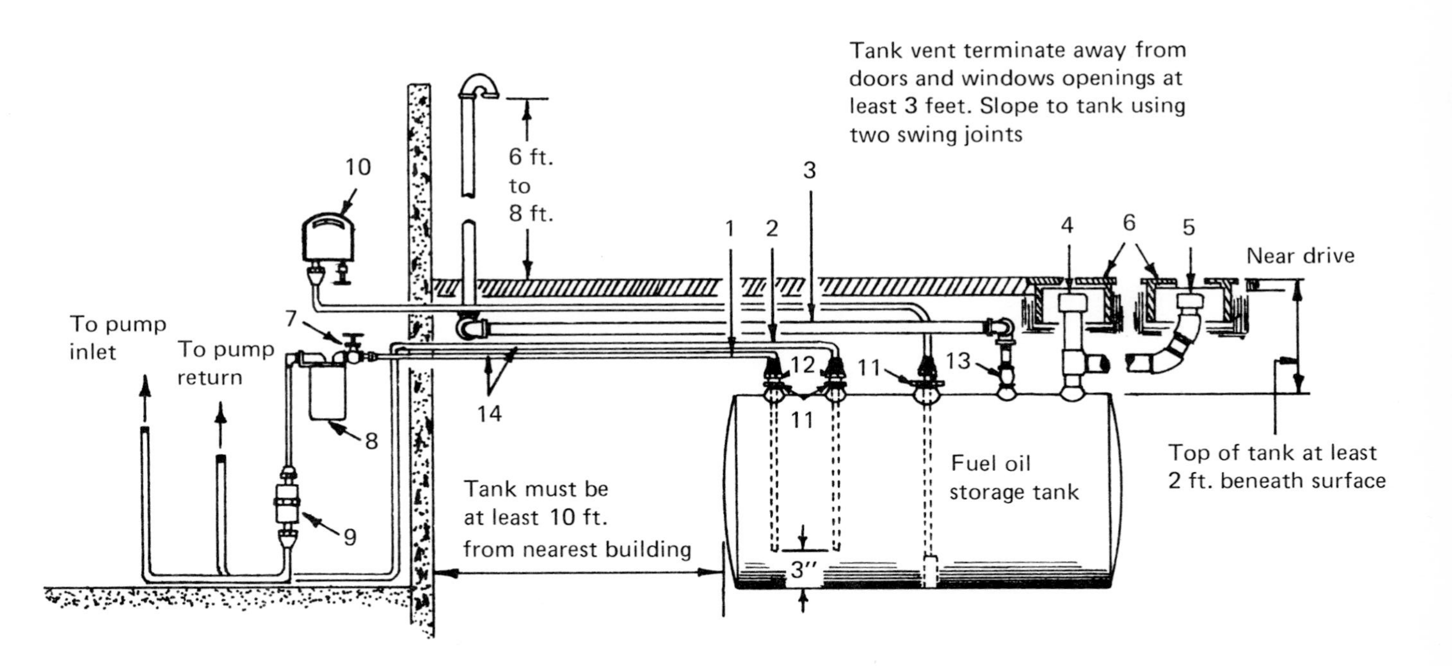

1. Suction line . . . . 3/8″ copper tubing
2. Return line . . . . 3/8″ copper tubing
3. Vent pipe . . . . 1 1/4″ galv. iron pipe
4. Test well . . . . 2″ galv. iron pipe
5. Fill pipe . . . . 2″ galv. iron pipe
6. Flush boxes
7. Shut-off valve . . . . 3/8″ fuse valve
8. Oil filter
9. Ball check valve
10. Fuel oil storage gauge
11. Reducing bushings
12. Slip fittings . . . . for 3/8″ copper tube
13. Ventalarm
14. Suction and return continuous runs. Avoid sharp bends

Figure 1.13. Outside tank installation.

Chapter 2
PRINCIPLES OF COMBUSTION

Combustion Is "Burning"

Burning is rapid oxidation. Oxidation is the combination of oxygen with any other substance. When this combination takes place, heat is released. The oxygen necessary for combustion is taken from the air. Air contains approximately 21% of oxygen. What actually occurs during combustion is as follows:

A solid or liquid fuel—whether coal or oil—is heated from an outside source such as a match, or electric spark, to a point where it produces a gas (vapor). This gas mixes with the air. If there is sufficient heat, the oxygen in the air combines with the gas released by the solid or liquid fuel and we have combustion, which will continue to burn.

For efficient combustion, the gas or vapor of the fuel oil must be burned as completely as possible. Only by so doing can we take advantage of the full heat content of fuel oil.

The average fuel oil—#2—used for home heating, normally contains 140,000 BTU's per gallon. A BTU—British Thermal Unit—is the standard measurement for heating value. A BTU is the amount of heat required to raise the temperature of one pound of water one degree F.

To secure the maximum heat value of a fuel oil, we must release all the BTU's. These five conditions are necessary to accomplish this:

1. **The Oil Must Be Changed into a Vapor**

Since the oil in liquid form does not burn, it must be changed into a vapor by sufficiently high **temperature.**

Atomizing burners prepare the oil for rapid vaporization by breaking it up first into a fine mist. This may be done by pressure.

The mist consists of millions of droplets of oil. These droplets expose a large amount of oil surface to the action of heat and air. The oil is vaporized from the surface of these droplets. The droplets decrease in size as the surface is evaporated until all the oil has been vaporized and burned. Vaporizing burners depend mainly upon heat.

2. **The Vapor Must Be Intimately Mixed with Air**

Although heat alone will vaporize the oil, no combustion could take place if air were not present to supply the oxygen. Sufficient air must be provided to furnish the oxygen required for combustion. In practice, about 2,000 cubic feet or 15 pounds of air is required to burn each gallon of oil, or 12 cubic feet for each 1,000 BTU. The burner should have 15 cubic feet, for proper burning and "***draft.***"

The efficiency of combustion depends on how complete the thorough air is mixed with the vapor and how much air is supplied. The method of ejecting air and oil from the burner creates **turbulance** by swirling. The swirling air mixes thoroughly with the atomized and vaporized oil.

In a pressure atomizing burner, the **turbulance** is started by the vanes or blades in the air tube. It is increased by having the oil spray and air rotate in opposite directions. It is further aided by the design and location of the combustion chamber. (If the air intake adjustment is shut down too much, poor **turbulance** may result because the velocity of the air is insufficient.)

3. **The Mixture Must Be Ignited**

When the oil burner starts, the combustion chamber is cool. There is no heat present to vaporize the oil. It is the job of the transformer to generate a electric spark, to heat a very small portion of the oil or oil mist. As the **temperature** rises, vaporization and ignition takes place.

This small portion of burning vapor then heats the surrounding vapor. The action is progressive and rapid until complete combustion is established.

4. **The Mixture Must Burn in the Presence of a Refractory**

After combustion has been started by the transformer, it must be maintained. This is accomplished by the heat from the flame itself and by the heat from the refractory of the combustion chamber.

A refractory is a material which will not readily burn or deteriorate at high temperature. It may be fire-brick, fire clay or heat resisting metal.

The refractory is raised to a **temperature** well above the final vaporizing temperature of the oil. This refractory reflects heat into all parts of the flame to provide a high degree of **temperature** to all of the burning oil-air mixture. Complete combustion should occur during the **time** this mixture is in the combustion chamber.

Temperature and the **time** of combustion are, therefore, of great importance.

Without the refractory, sufficient heat would not be available to vaporize and burn the fuel completely. Smoke and Carbon would result.

The **time** of combustion is controlled to a great extent by varying the air supply, and by draft adjustment. The time of combustion is also affected by flame shape and characteristics and by combustion chamber size, design and location.

5. **The Products of Combustion Must Be Carried Away from the Flame***

This is accomplished by means of draft which carries the products of combustion away from the flame and up the chimney. If the product of combustion were permitted to accumulate around the flame they would smother the fire. The final products of complete combustion are carbon dioxide (CO_2) and water vapor (H_2O). In addition, nitrogen from the air is also present although it is not a product of combustion.

The products of combustion are formed chemically through rapid oxidation. Air, as we already know, is composed of about 21% of oxygen, 78% nitrogen, with 1% of other gases. The average fuel oil of the No. 2 grade is composed of about 85% carbon and about 12% hydrogen, with 3% of other elements.

When fuel oil is completely burned, all the carbon in it combines with oxygen to form carbon dioxide or CO_2. All the hydrogen combines with oxygen to form water vapor, or H_2O. The nitrogen in the air does not combine with anything. It passes off chemically unchanged.

*(In discussing the products of combustion we are referring to the products which exist at the smoke pipe. There are many intermediate reactions involved in burning the carbon and hydrogen in fuel oil, but only the final gases produced need be considered.)

Under this condition all the BTU's in the oil are released. This is possible (only theoretically) when there is sufficient air to furnish enough oxygen for complete combustion.

If the air supply is insufficient, here is what will happen—the hydrogen, because it combines more readily with oxygen, will usually be converted entirely to water vapor (H_2O). But all the carbon will not be changed to carbon dioxide (CO_2). Some of it will become carbon monoxide (CO) and burn. This will produce only about one-third as much heat as when the carbon becomes carbon dioxide (CO_2). And some of the carbon, moreover, may not burn at all but will pass off as smoke, or be deposited as soot. This carbon, because it is unburned, releases no heat. Obviously, we want to convert as much of the carbon in the fuel as possible to carbon dioxide (CO_2).

If we were concerned only with complete combustion, we could supply an unlimited amount of air to the combustion chamber—or, rather, as much as the chimney could carry off. But efficient combustion demands not only complete burning but also the most efficient application of the released heat to the heat absorbing surfaces of the furnace.

For this reason, it is advantageous to burn the fuel as completely as possible with the minimum supply of air. If we have too much air, we are using fuel to heat this excess air from the low temperature of the basement to a temperature from five to ten times higher.

If no excess air were present and still the fuel oil were completely burned, as is possible only in the laboratory, the products of combustion would contain 15% CO_2—no CO—and no smoke. The percentage of CO_2 and CO, and the presence or absence of smoke, therefore, may be used to indicate how completely and efficiently the fuel is burned.

In practice, some excess air is necessary to secure complete combustion. Depending upon the type of burner and the operating conditions of the installation, the maximum practical CO_2 percentage will range from 9 to 12, corresponding to 68 to 26 percent excess air. The excess air determines the total volume of gas discharged to the smoke pipe. As this volume is increased by increasing the amount of excess air, the percentage of CO_2 decreases, although the actual amount of CO_2 remains constant. This condition enables us to determine percent of excess air from the CO_2 percentage.

It should be kept in mind that CO_2 percentage can always be raised by decreasing excess air—either by closing up on the air shutter or by sealing air leaks. If, however, such an adjustment results in a smoking fire, it will not be satisfactory. In practice it is a safe rule to set a burner at one percent less CO_2 than the percent at which smoke first becomes detectable by the smoke test. If this point is too low in CO_2 for good efficiency, a cause for poor combustion (too much or too little draft, unsuitable combustion chamber, dirty nozzle, burner incorrectly located, etc.) should be found and corrected.

Troubles, Complaints and Remedies

We have seen that combustion is really a chemical process. What we are concerned with here is the degree of combustion efficiency.

Combustion efficiency is determined:

1. By an analysis of the products of combustion.
2. By checking the temperature of these products of combustion as they leave the furnace.
3. By measuring the amount of draft.

Analysis of the Products of Combustion

A "CO_2 Indicator" is used to make this analysis. There are two types in general use. One determines the percentage of CO_2 chemically. The other does it electrically.

Stack Temperature

Stack temperature readings are taken principally to determine how much of the heat produced during combustion is absorbed by the furnace.

Stack temperatures from 400° to 700° Fahrenheit are usually found in practice. Less than 350°F may cause condensation of water vapor in the smoke pipe and chimney. More than 550°F indicates that useful heat is escaping before it can be absorbed by the furnace. This is known as "Stack Loss."

Stack temperature and CO_2 readings are related. This relationship is seen from the Fuel Loss Chart, which indicates combustion efficiency based on the carbon dioxide (CO_2) and stack temperature readings.

Both of these readings should be taken at, or close to, the smoke pipe connection to the furnace. They should be taken after the burner has been in operation for about ten or fifteen minutes.

The bulb of the thermometer and the end of the tube through which the gases are drawn to the CO_2 indicator should be at the center of the pipe.

Some furnaces provide an opening into the thimble, in others a hole must be drilled. If air leaks in through this opening, it will make an accurate reading impossible. Asbestos should be used around the tube and thermometer in order to prevent this air leakage.

TEST YOUR KNOWLEDGE

1. What is the heating value of # 2 Fuel Oil in BTU's. ____________________

2. Air contains approximately what percent of oxygen ____________________

   ____________________ .

3. The three requirements for combustion to take place are liquid, oil, air, and ignition (T) or (F).

4. For good combustion you should have about 12 cubic feet for each 1,000 BTU's (T) or (F)

5. The final products of complete combustion would be.
   A. SO_2 H_2O
   B. CO_2 H_2O
   C. SO_2 NH_3
   D. CO_2NH_3

6. The basic constituents of air would be.
   A. 78% Nitrogen 21% Oxygen 1% Other Gas
   B. 21% Nitrogen 78% Oxygen 1% Other Gas
   C. 1% Nitrogen 78% Oxygen 21% Other Gas
   D. 1% Oxygen 78% Nitrogen 21% Other Gas

7. The basic constituents of #2 Fuel Oil would be about ____________________ .
   A. 21% Oxygen 78% Nitrogen 1% ____________
   B. 78% Oxygen 21% Nitrogen 1% ____________
   C. 85% Carbon 12% Hydrogen 3% ____________
   D. 12% Carbon 85% Hydrogen 3% ____________

8. Stack temperatures should be no less than 350 °F or more than 550 °F ____________ .
   (T) or (F)

9. The two types of CO_2 indicators would be *chemically* and *electrically* ____________ .
   (T) or (F)

10. The conditions that are necessary to accomplish good combustion would be
    1. Change the oil into a vapor.
    2. The vapor must be mixed with air.
    3. The mixture must be ignited.
    4. The mixture must be burned in the presence of a refractory.

    (T) or (F)

STARTING THE OIL BURNER

IMPORTANT CAUTIONS: Before Starting

1. *Fuel Unit*

Be sure that the fuel unit is arranged for the "one pipe or two pipe system"; be sure that all connections are tight. All single stage pumps require manual venting, regardless of whether connected for one pipe or two pipe systems.

Two stage fuel units with two pipe systems only are self-venting, through the return line.

2. *Nozzle*

Be sure that the specified nozzle is installed, and that any covering over the nozzle, that is placed there to keep it clean during installation, is removed.

3. *Electrodes*

Check electrode spacings and adjustments with respect to the nozzle See Figure 1.6, Chapter 1.

4. *Oil*

Be sure that there is oil in the tank, that the tank is provided with a proper vent pipe. See tank installation Figure 12, Chapter 1.

5. *Power*

Be sure that the voltage of the power supply conforms to burner requirements (115V-60C); that the burner and controls are wired correctly and that the line switch is properly fused 15 amp.

6. Adjust the limit control to the desired setting.

Starting Procedure

1. Set the thermostat adjustment above room temperature.
2. Open the manual shut-off valve in the oil supply line to the burner.
3. Check initial air control adjustment should be about one half open.
4. Close line switch, to start the burner. If the burner does not start simultaneously with the line switch, it will likely be because the contacts of the stack control are not in the "cold" or starting position, see stack controls.
5. Vent the fuel unit as soon as the burner starts. (Single stage units with one or two pipe systems, or two stage units with one pipe system Venting Procedure: Loosen the vent plug, do not remove completely, only about half way out. Hold an empty can under the vent opening to catch the oil which will be expelled from the unit, as soon as the air has been expelled. Drain at least 1/4 pint of oil from the unit, as soon as the fuel unit is vented, the pressure regulator will allow opening of the nozzle outlet valve, and oil will be supplied to the nozzle under regulated pressure. Ignition should be instantaneous. If the burner starts and runs but stops again during the "venting operation" or a few seconds after, wait three to five minutes for the safety switch of the stack control to cool, then reset the safety switch. Two stage fuel units with two pipe systems only are self venting through the return line.

6. *Air Adjustment*

Adjust the air supply to allow just sufficient air for clean combustion. Reduce the air supply until the flame tips appear slightly smokey, then increase just enough to make the flame tips appear absolutely clean.

7. *Draft Control Adjustment*

Adjust the draft control for minimum draft that will carry away the products of combustion completely. In most installations the control may be adjusted for minimum draft, that is in the balanced position full open, and with the burner running and fire door closed. When the burner air supply and draft will be −.02″ to −.03″ w.c., larger installations may require slightly higher draft.

8. *Pressure Adjustment*

All fuel units must be checked to see if they are in the 95 to 105 lb. range, most fuel units come from the factory adjusted to 100 lbs. of pressure. At this pressure, with No. 2 fuel oil, the oil rate should be within 5% of the nozzle rating and with most oils atomization should be sufficient for clean, quiet combustion. Occasionally, it may be desirable to raise or lower the atomizing pressure. Raising the pressure increases the intensity of the flame, by increasing the rate of burning and finer atomization of oil. Lowering the pressure has the opposite effect. The gauge used for pressure tests should be from (0 to 200) p.s.i.g.

9. *Check Air Adjustment*

After the draft control has been adjusted and locked, and the oil pressure has been adjusted if necessary, allow the burner to run about 10 minutes, then check the air adjustment, lock the air shutter adjustment before leaving the burner.

10. *Flame Setting*

The body of the flame with correct oil rate, pressure, draft and burner air adjustment should be uniform and clean, with the flame tips just off haze. Any further reduction of air supply should tend to elongate the flame tips and make them smokey. Combustion should be complete, or practically so, within the combustion chamber. If there is the slightest indication of pulsation at starting, then a slight amount of secondary air should be provided at the fire door. The stack draft should be just sufficient to carry away the products of combustion (approx. −.02″ w.c.) and the CO_2 over the fire, or in the fire gas should be not less than 8%. See combustion testing Chapter 3 for final flame setting.

11. *Checking the Controls*

Check and adjust all controls in accordance with the manufacturers' instruction sheets, (fan and limit controls). Before attempting to check operation of the combustion control, be sure that the burner flame has been properly adjusted, (stack relay or cadcell). If the furnace is provided with a manual reset limit control, normally included with counterflow and horizontal units, particular attention should be paid to the special instructions as regards control adjustment.

12. *Other Checks*

Be sure the air shutter and draft control adjustments are locked, that there is an ample supply of fresh air to the furnace room, that all controls are properly installed and adjusted; that line fuses are tight and there are no oil leaks.

13. *Lubrication*

Most oil burner motors have oil cups, oil with a few drops of No. 10 motor oil nondetergent. If the installation is equipped with a blower motor or circulator, oil with No. 20 motor oil nondetergent only.

14. *Operating Instructions*

Instruct the owner or responsible member of the household, the care and operation of the heating system should be explained to the home owner; also how to adjust the thermostat, necessity of air supply to the burner, and the following simple checks to make before calling for service. If the burner fails to operate automatically:

1. Be sure there is oil in the tank.
2. Be sure the thermostat is set above room temperature.
3. Be sure the main line switch is on, and the fuses are OK.

15. *Follow-up Inspection*

After the burner has been in operation a few days, it should be inspected to check:

1. Flame adjustment
2. Automatic starting and stopping
3. Limit control adjustment and operation
4. Thermostat operation and adjustment
5. Possibility of oil leaks
6. Uniformity of heating

This will show the owner that you are interested in his heating unit and not just his money.

TEST YOUR KNOWLEDGE

1. Be sure that the fuel unit is arranged for the, one pipe or two pipe system's. (T) or (F)
2. Be sure that the specified nozzle is installed. (T) or (F)
3. Be sure that the air adjustment is all the way open. (T) or (F)
4. The pressure of the fuel unit should be set for ________________ .
   A. 90#
   B. 100#
   C. 110#
   D. 120#
5. The stack draft should be set for ________________ .
   A. −.2″W.C.
   B. −.02″W.C.
   C. +.03″W.C.
   D. +.3″W.C.
6. Be sure that the oil tank valve is off, before starting the oil burner. (T) or (F)
7. Be sure that the voltage and fuse are properly sized, *115V AC* 20 AMP. Fuse. (T) or (F)
8. Be sure the air shutter and draft control adjustments are locked when done with the job. (T) or (F)
9. Allow the burner to run about one minute then check the air adjustment. (T) or (F)
10. Oil the burner motor with SAE#20 motor oil detergent. (T) or (F)

Chapter 3
COMBUSTION TESTING

Common Causes of Low CO_2 and Smoke Fire on Oil Burners

1. Improper fan delivery or incorrect air shutter opening.
2. Nozzle is worn clogged or of incorrect type.
3. Electrodes are dirty, loose or incorrectly set.
4. Draft regulator is improperly installed or sticking.
5. Oil pressure to nozzle improperly adjusted causing poor spray characteristics.
6. Nozzle is loose or not centered.
7. GPH rate is too high for size of combustion chamber.
8. Spray angle of nozzle unsuited to air pattern of burner or shape of firebox.
9. Air handling parts defective or incorrectly adjusted.
10. Firebox is cracked or of improper refractory material.
11. Nozzle spray or capacity unsuited to the particular type of burner being used.
12. Furnace or boiler has excessive air leaks.
13. Burner "on" periods are too short.
14. Cut-off valve leaks allowing after-drip of fuel oil.
15. Ignition is delayed due to defective stack control.
16. Oil does not conform to burner requirements.
17. Draft is insufficient due to defective flue or insufficient height of chimney.
18. Rotary burner motor is running underspeed.

COMBUSTION TESTING WITH BACHARACH

Stack Thermometer

The Bacharach stack thermometer is an instrument which in the heating industry is used to determine the condition of heating equipment. It is used in conjunction with other combustion-testing instruments to diagnose and solve afflictions of the combustion process. To use the stack thermometer drill a 1/4 inch hole in the flue pipe. This hole should be about 12 inches from the furnace or boiler breeching or outlet and on the furnace side of the draft regulator at least 6 inches away from it. Turn the oil burner on and allow it to operate about ten minutes before beginning the test. Insert the thermometer stem into the test hole. (If it is of the clip-holding type, use the clip to secure it to the flue.) Read the temperature on the dial scale at which the pointer finally comes to rest. Determine the net stack temperature by subtracting the basement air temperature from the thermometer reading. Thermometer reading 600 °F basement temperature 70 °F. Net stack temperature 600 °F minus 70 °F = 530 °F for that unit. As a rule of thumb, stack temperature reading in excess of the following figures may generally be considered abnormally high. Such high readings are caused for special concern and ad-

Figure 3.1. Bacharach stack thermometer.

Figure 3.2. Bacharach CO_2 tester.

justments. Consult the manufacturer's specification sheets for individual units. Conversion units—600 °F to 700 °F. Packaged units 350 °F to 550 °F. A high stack temperature may indicate any of the following conditions. These should be immediately checked and remedied if the target is peak efficiency. (1) Excessive draft through the furnace or boiler. (2) Dirty, carbon covered furnace or boiler heating furnaces. (3) Lack of sufficient baffling. (4) Undersized furnace. (5) Incorrect or defective combustion chamber. (6) Furnace or boiler overfired. (7) Improper adjustment of draft regulator.

Bacharach CO_2 Indicator

There are five simple steps which you must take to insure an accurate CO_2 test with the Bacharach CO_2 indicator. (1) For taking the flue gas sample, use the hole in the flue pipe previously drilled for the stack thermometer. (2) Turn the oil burner on and allow it to operate about ten minutes before running the test. (3) Insert the sampling tube of the CO_2 indicator's gas aspirating assembly into the hold in the flue pipe. The rubber cap end is placed on the top, or plunger valve, of the indicator and held in a depressed position. The aspirator rubber bulb is next squeezed 18 times in succession. On the 18th squeeze the depressed plunger valve is released before releasing the rubber bulb. (4) The indicator is now turned over twice, permitting the test fluid to run back and forth, and forcing it to absorb the flue gas sample. This turn- over motion is the same action one might make with an hour-glass or egg-timer. (5) The indicator is placed or held in an upright, level position and the test liquid is read on the scale which is calibrated directly in percent of CO_2. The highest possible percent CO_2 reading should always be the goal (with satisfactory smoke rating considered). When simple adjustments are possible, 10% CO_2 (with Bacharach smoke scale reading No. 2) may be considered satisfactory. If major adjustments must be made, the following rules may be used. Where the net stack temperature is less than 400 °F a CO_2 reading of as low as 8% may be tolerated. Where the net stack temperature is over 500 °F, a CO_2 reading of at least 9% should be the target. A low percent CO_2 reading is indicative of any of the following conditions which should be remedied. (1) High draft. (2) Excess combustion air. (3) Incorrect or defective firebox. (4) Air leakage. (5) Poor oil atomization. (6) Worn, plugged or incorrect nozzle. (7) Furnace or boiler has excessive air leaks. (8) Incorrect air-handling parts. (9) Draft regulator operates erratically. (10) Oil pressure is yet incorrectly. If too great a difference in CO_2 readings occurs between samples taken in the flue pipe and through the fire door, an air leakage or other unsatisfactory combustion condition exists in the furnace or boiler.

Bacharach Combustion Efficiency Slide Rule

Two factors determine heat lost in the flue gases, percent of CO_2 in the flue gases and stack temperature. The flue gas heat loss in turn determines the combustion efficiency of the heating plant. The combustion efficiency slide rule provides a rapid, simple means to determine combustion efficiency and stack loss from the results of the CO_2 and stack temperature test. The slide rule has horizontal and vertical slide inserts. The horizontal slide is moved until the measured stack temperature appears in the window marked, stack temperature. Next, the vertical slide is moved until the black arrow points to the measured CO_2 percent combustion efficiency and stack loss are then indicated in the cut-out of the arrow on the vertical slide.

Bacharach Smoke Tester

The objective of the smoke test is to measure the smoke content in the flue gases and then, in conjunction with other combustion test results, adjust the burner to optimum operation. The smoke scale which is always used in conjunction with the smoke tester, has ten color graded spots from (o)-pure white to (9) the darkest color. After the burner has been in operation for about ten minutes, insert the free end of the tester's sampling tube into the hole in the flue pipe used for the two previous tests, place the filter paper into the holding slot of the tester, pull the smoke tester handle through ten full pump strokes, holding for several seconds between each pumping stroke. Remove the filter paper from with the smoke scale numbers on the scale, matching your sample with the closest color on the smoke scale. It should be emphasized also that not all types of oil burners will be equally affected by the same smoke content in the flue

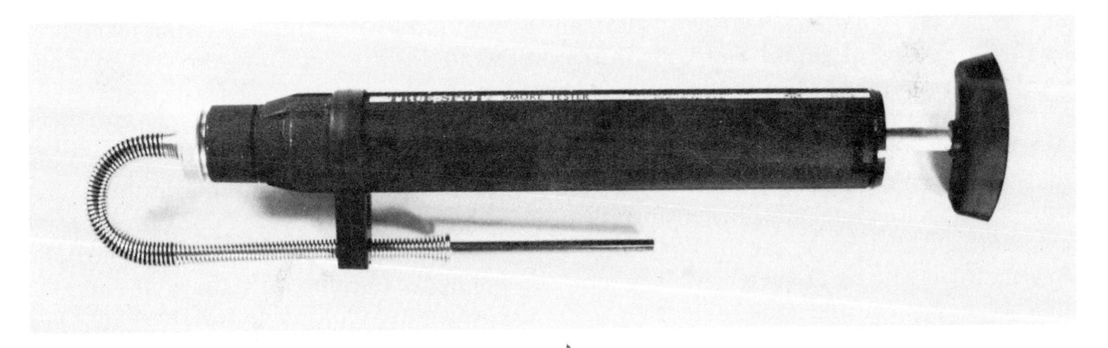

Figure 3.3. Bacharach smoke tester.

gas. This is shown in the table which interprets smoke scale readings in terms of sooting produced. Depending upon the construction of the heat exchanger or the boiler, some units will accumulate soot rapidly at a No. 3 smoke, while the accumulation of soot on other units at the same smoke scale reading may be relatively slower.

Smoke Scale No.	Rating	Sooting Produced
1	Excellent	Extremely light if at all
2	Good	Slight sooting which will not increase stack temp. appreciably
3	Fair	May be some sooting but will require cleaning once a year
4	Poor	Some units will require cleaning more than once a year
5	Very Poor	Soot rapidly and heavily

Smoky combustion can be (1) Improper fan delivery, (2) Insufficient draft (3) Poor fuel supply (4) Oil pump not functioning properly (5) Nozzle defective or incorrect type (6) Furnace or boiler has excessive air leaks (7) Improper fuel-air ratio (8) Fire box defective (9) Draft regulator improperly adjusted (10) Improper burner air handling parts.

Bacharach Draft Gauge

Correct draft is essential for efficient burner operation. While draft is not directly related to combustion efficiency, it affects oil burner efficiency. The intensity of the draft determines the rate at which combustion gases pass through the boiler or furnace. The intensity of draft also governs the amount of air supplied for combustion. Excessive draft can increase the stack temperature and reduce the percent CO_2 in the flue gases. The draft requirement varies with each type of installation. Insufficient draft may cause pressure in the combustion chamber, leading to escape of smoke and odor in the basement. Insufficient draft also makes it impossible to adjust the burner for highest efficiency, because maximum efficiency depends on the proper amount of air mixing with the correct amount of oil each time the burner runs. Drill a 1/4 inch hole in the fire door or if feasible remove one of the bolts holding the fire door lining to the door. This hold is used for the overfire draft reading. For the flue pipe the draft reading, use the hole in the flue pipe previously drilled for the other tests. Place the draft gauge on any convenient level surface near the boiler or furnace. Adjust the draft gauge to zero. Allow the furnace to operate for about ten minutes before beginning the test. Insert the draft tube into the test holes and read the gauge. These two readings which are indicative of draft conditions. Overfire draft reading and flue draft reading.

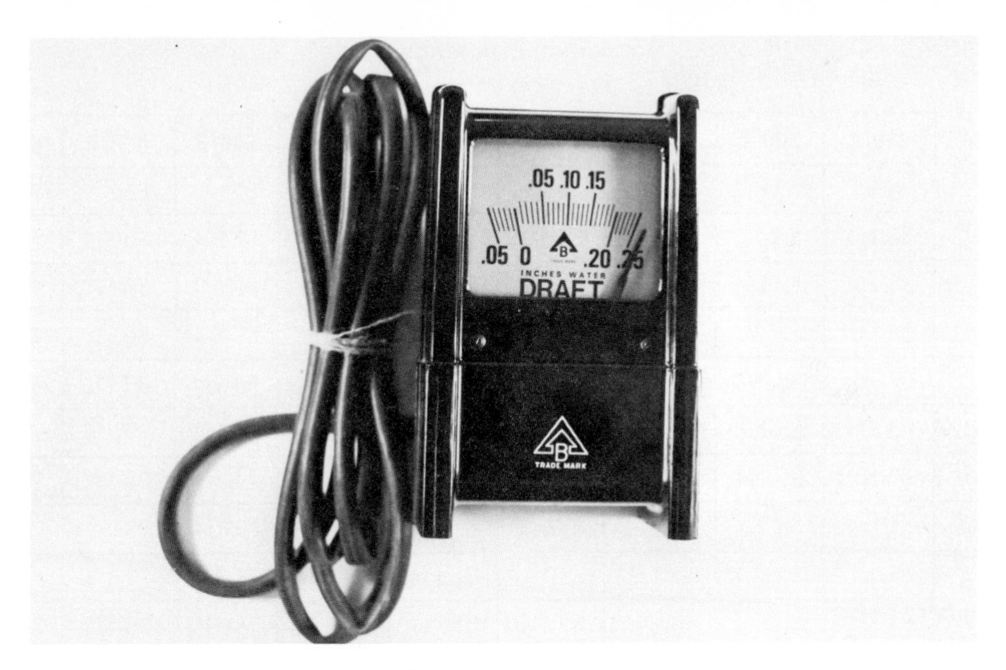

Figure 3.4. Bacharach draft gauge.

Overfire draft of not less than -0.01 inches water may be considered sufficient to develop and maintain proper combustion. If the firebox (overfire) draft runs less than -0.01 inches water, smoke and odor may occur in the basement. Also rumblings may occur if close settings for highest efficiencies are used. Flue pipe draft must be sufficient to prevent positive pressure in the combustion chamber. A furnace having long and complex flue passages requires a higher draft in the smoke pipe than a furnace with short flue passages. Most residential oil burners with firing rates not above 1.5 GPH need flue pipe draft between -0.03 to -0.04 inches water in order to maintain a draft of -0.01 inches water in the firebox.

Bacharach Efficiency Oil

CO_2%	250°F	300°F	350°F	400°F	450°F	500°F	550°F	600°F	650°F
15	88 1/2	87 1/2	86 1/2	85 1/4	84 1/4	83 1/4	82	81	79 3/4
14 1/2	88 1/4	87 1/2	86 1/4	85	84	83	81 3/4	80 3/4	79 1/4
14	88 1/4	87 1/4	86	84 3/4	83 3/4	82 3/4	81 1/2	80 1/4	79
13 1/2	88	87	85 3/4	84 1/2	83 1/2	82 1/2	81 1/4	80	78 3/4
13	88	86 3/4	85 1/2	84 1/4	83 1/4	82	80 3/4	79 1/2	78 1/4
12 1/2	87 3/4	86 1/2	85 1/4	84	83	81 1/2	80 1/4	79	77 3/4
12	87 1/2	86 1/4	85	83 3/4	82 1/2	81 1/4	79 3/4	78 1/2	77 1/4
11 1/2	87 1/4	86	84	83 1/2	82	80 3/4	79 1/4	78	76 1/2
11	87	85 3/4	84 1/2	83	81 1/2	80 1/4	78 3/4	77 1/4	75 3/4
10 1/2	86 3/4	85 1/2	84	82 1/2	81	79 1/2	78	76 1/2	75
10	86 1/2	85	83 1/2	82	80 1/2	78 3/4	77 1/4	75 3/4	74 1/4
9 1/2	86 1/4	84 1/2	83	81 1/2	79 3/4	78	76 1/2	75	73 1/4
9	85 3/4	84	82 1/4	80 3/4	79	77 1/4	75 3/4	74	72 1/4
8 1/2	85 1/4	83 1/2	81 3/4	80	78 1/4	76 1/2	74 3/4	73	71 1/4
8	84 3/4	83	81	79 1/4	77 1/2	75 1/2	73 3/4	71 3/4	70
7 1/2	84 1/4	82 1/4	80 1/4	78 1/2	76 1/2	74 1/2	72 1/2	70 1/2	68 1/2
7	83 3/4	81 1/2	79 1/2	77 1/4	75 1/4	73 1/4	71	69	67
6 1/2	83	80 3/4	78 1/2	76 1/4	74	71 3/4	69 1/2	67 3/4	65
6	82 1/4	79 3/4	77 1/4	75	72 1/2	70	67 3/4	65 1/4	62 3/4
5 1/2	81 1/4	78 1/2	76	73 1/2	71	68	65 1/2	63	60 1/4
5	80	77 1/4	74 1/2	71 3/4	69	65 3/4	63	60	57
4 1/2	78 1/2	75 1/2	72 1/2	69 1/2	66 1/4	63	60	56 3/4	53 1/2
4	76 3/4	73 1/4	69 3/4	66 1/4	62 3/4	59 1/4	55 3/4	52	48 1/2

Example of combustion efficiency. With a stack temperature of 550°F and a CO_2 of 8 1/2%, the unit would be 74 3/4% efficient.

Bacharach Efficiency Natural Gas

CO_2%	250°F	300°F	350°F	400°F	450°F	500°F	550°F	600°F
11	85 1/4	84	82 3/4	81 3/4	80 3/4	79 3/4	78 1/2	77 1/2
10 1/2	84 3/4	83 1/2	82 1/2	81 1/2	80 1/2	79 1/4	78	77
10	84 1/2	83 1/4	82 1/4	81	80	78 1/2	77 1/2	76
9 1/2	84 1/4	83	81 3/4	80 3/4	79 1/2	78 1/4	77	75 3/4
9	84	82 1/2	81 1/4	80 1/4	78 3/4	77 1/4	76	74 3/4
8 1/2	83 1/2	82 1/4	81	79 3/4	78 1/4	77	75 1/2	74
8	83 1/4	81 3/4	80 1/2	79	77 1/2	76	74 1/2	73
7 1/2	82 3/4	81 1/4	80	78 1/4	76 1/2	75 1/4	73 1/2	72
7	82 1/4	80 3/4	79	77 1/2	75 3/4	74 1/4	72 1/2	71 1/4
6 1/2	81 3/4	80 1/4	78 1/4	76 1/2	75	73	71 1/2	70
6	81 1/4	79 1/2	77 1/4	75 1/2	73 1/2	71 3/4	70 1/4	68 1/2
5 1/2	80 1/2	87 1/4	76 1/4	74	72 1/4	70 1/2	68 1/2	66 3/4
5	79 1/2	77	75	72 1/2	70 3/4	68 3/4	66 1/2	64 1/2
4 1/2	78 1/4	75 3/4	73 1/4	71	69	66 1/2	64	62
4	76 3/4	74	71 1/2	69	66 1/2	64	61 1/4	59

Example of combustion efficiency. With a stack temperature of 350°F and a CO_2 of 5%, the unit would be 75% efficient.

Bacharach Efficiency Propane Gas

CO_2%	250°F	300°F	350°F	400°F	450°F	500°F	550°F	600°F
13 1/2	87 1/4	86 1/2	85 1/2	84 1/2	83 1/2	82 3/4	81 3/4	80 3/4
13	87 1/4	86 1/4	85 1/4	84 1/4	83 1/4	82 1/4	81 1/4	80 1/4
12 1/2	87	86	85	84	83	82	81	80
12	86 3/4	85 3/4	84 3/4	83 3/4	82 3/4	81 1/2	80 1/2	79 1/2
11 1/2	86 1/2	85 1/2	84 1/2	83 1/4	82 1/4	81	80	79
11	86 1/4	85 1/4	84 1/4	83	81 3/4	80 3/4	79 3/4	78 1/2
10 1/2	86 1/4	85	83 3/4	82 1/2	81 1/2	80 1/4	79	78
10	86	84 3/4	83 1/2	82 1/4	81	79 3/4	78 1/2	77 1/4
9 1/2	85 1/2	84 1/4	83	81 3/4	80 1/2	79	77 3/4	76 1/2
9	85 1/4	84	82 1/2	81 1/4	79 3/4	78 1/2	77	75 3/4
8 1/2	84 3/4	83 1/2	82	80 1/2	79 1/4	77 3/4	76 1/4	74 3/4
8	84 1/2	83	81 1/4	79 3/4	78 1/4	76 3/4	75 1/4	73 3/4
7 1/2	83 3/4	82 1/4	80 1/2	79	77 1/2	75 3/4	74 1/4	72 1/2
7	83 1/2	81 3/4	80	78 1/4	76 1/2	74 3/4	73	71 1/4
6 1/2	82 3/4	81	79	77 1/4	75 1/4	73 1/2	71 3/4	69 3/4
6	82	80	78	76 1/4	74 1/4	72 1/4	70 1/4	68 1/4
5 1/2	81 1/4	79 1/4	77	75	72 3/4	70 3/4	68 1/2	66 1/4
5	80 1/4	78	75 1/2	73 1/4	70 3/4	68 1/2	66	66 3/4
4 1/2	79	76 1/2	74	71 1/4	68 3/4	66	63 1/2	60 3/4
4	77 3/4	74 3/4	71 3/4	69	66	63	60 1/4	57 1/4

Example of combustion efficiency. With a stack temperature of 500°F and a CO_2 of 9 1/2%. The unit would be 79% efficient.

Dwyer CO_2 Indicator No. 1101

Operating Instructions

1. Depress inlet and vent plungers. Release both slowly and set zero mark on the scale at the top of the liquid column in the indicating tube.
2. Insert sampling tube of rubber tubing assembly in smoke pipe between draft control and furnace. Place inlet valve connector on the inlet plunger and depress. *Pump 20 full slow strokes of the bulb* and release the inlet plunger with the bulb fully compressed from the last stroke.
3. Tip indicator slowly back and up again several times. *Do not invert.*
4. Read percentage of CO_2. For subsequent tests, it is unnecessary to reset the zero as in step (1). Simply follow the procedure in steps (2) and (3). For greatest accuracy the indicator should be at room temperature. If it is cold, readings will be slightly high. Should the solution fall below the "fluid level" mark carefully add a few drops of tap water to bring the level up to the proper point.

Recharging Dwyer CO_2 Indicator

1. After 500 tests or one year, the absorbent solution should be replaced.
2. Remove both valve assemblies, pour out the old solution and clean the entire instrument thoroughly. Use only clear water.
3. Refill with red absorbent solution to the fluid level mark.
4. Lubricate "0" ring seals slightly with vaseline or light oil and replace valve assemblies. Finger tight only.

Dwyer Inclined Draft Gage No. 171

Operating Instructions

1. Remove gage from case and swing table support perpendicular to the body of the gage. Place on reasonable level surface.
2. Turn connector clockwise 1 1/2 turns, thus venting to atmosphere.
3. Center bubble between cross hairs on spirit level with leveling screw.
4. Turn zero adjust screw so that meniscus of fluid lies directly in front of "0" mark on scale.
5. If there is insufficient oil to adjust the scale correctly, add oil a drop at a time from the 3/4 ounce bottle. If there is too much oil to adjust the scale, a pipe cleaner may be used to remove the excess.
6. For draft reading, connect the rubber tubing to the right side of the gage and place the terminal tube in the source of draft, either over the fire or in the smoke pipe on the furnace side of the draft control.
7. The draft may then be read directly on the scale and will be indicated in hundredths of an inch of water.

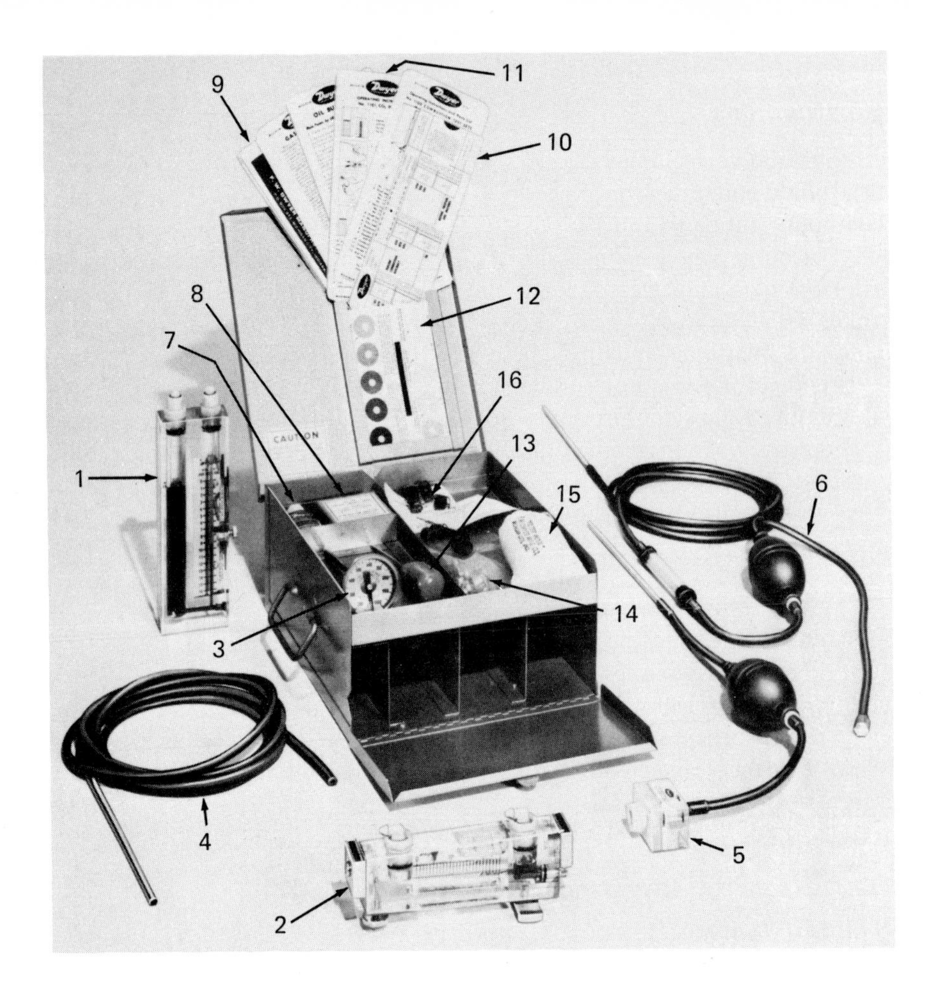

1. CO_2 Indicator
2. Draft gage
3. Stack thermometer
4. Draft gage tubing
5. Smoke gage and tubing assembly
6. CO_2 indicator bulb, filter and tubing assembly
7. Replacement draft gage fluid
8. Smoke sampling papers
9. Operating instructions and parts list
10. Combustion efficiency slide rule
11. Combustion data cards
12. Smoke chart
13. Awl to pierce smoke pipe for thermometer and sampling tube
14. Smoke pipe test hole plug pack
15. Bag—Filter wool
16. Terminal tube holder spring

Figure 3.5. Combustion testing with Dwyer test set no. 1100.

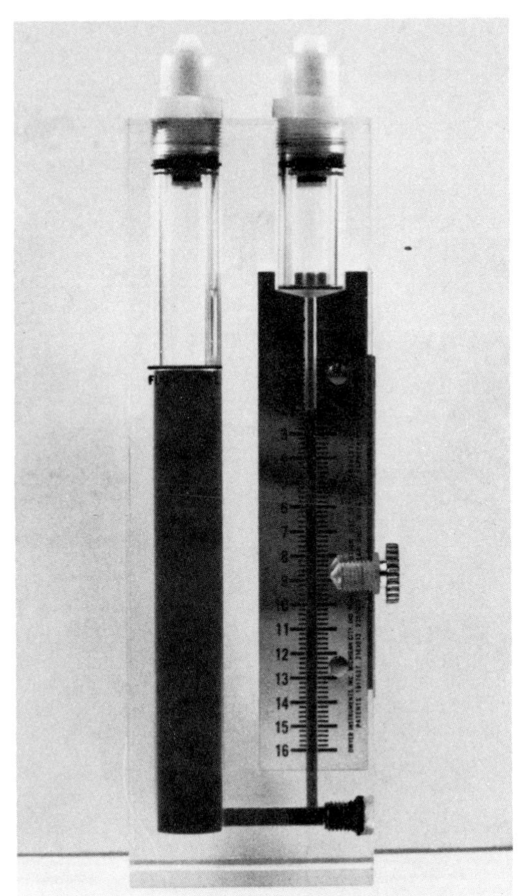

Figure 3.6. Dwyer CO_2 tester.

Figure 3.8. Dwyer smoke tester.

Figure 3.7. Dwyer inclined draft gage.

Dwyer Combustion Efficiency Oil

CO_2%	300°F	350°F	400°F	450°F	500°F	550°F	600°F	650°F
15	90	88	86	85	83 1/2	82	81	79 1/2
14	89	87 1/2	85 1/2	85	83	81 1/2	80 1/2	79
13	89	87	85	84 1/2	82	81	79 1/2	78
12	88	86 1/2	84 1/2	83 1/2	81 1/2	79 1/2	78	77
11	87 1/2	86	84	82	80 1/2	79	77	75
10	87	85	83	81	79 1/2	77	75	73
9	86	84 1/2	81 1/2	79 1/2	77 1/2	75	73	71 1/2
8	85 1/2	83	81	77 1/2	75 1/2	73 1/2	71	68 1/2
7	84 1/2	81 1/2	79	75 1/2	73 1/2	71	68	65
6	82	80 1/2	75 1/2	73 1/2	70	67	64	61 1/2
5	80 1/2	77	73	70	66	63	60	56

Example of combustion efficiency. With a stack temperature of 550°F and a CO_2 of 9%, the unit would be 75% efficient.

Dwyer Combustion Efficiency Natural Gas

CO_2%	300°F	350°F	400°F	450°F	500°F	550°F	600°F	650°F
11	82 1/2	81 1/2	80	79	78	77	76	75
10	82	81	79 1/2	78 1/2	77 1/2	76	75	74
9	81 1/2	80 1/2	79	78	76 1/2	75	74	72 1/2
8	81	79 1/2	78 1/2	77	75 1/2	74	72 1/2	71
7	80	78 1/2	77	75 1/2	74	72 1/2	71	69 1/2
6	79 1/2	77 1/2	75 1/2	74	72	70 1/2	68 1/2	67
5	78	76	74	71 1/2	69 1/2	67 1/2	65 1/2	63
4	76	73 1/2	71	68	65 1/2	63	60 1/2	58
3	72 1/2	69	65 1/2	62	59	56	52 1/2	51

Example of combustion efficiency. With a stack temperature of 400°F and a CO_2 of 10%, the unit would be 79 1/2% efficient.

Dwyer Combustion Efficiency Propane Gas

Co_2%	300°F	350°F	400°F	450°F	500°F	550°F	600°F	650°F
13	88	87	86	85	83 1/2	82 1/2	81 1/2	80 1/2
12	87 1/2	86 1/2	85 1/2	84	83	82	80 1/2	79 1/2
11	87 1/2	86	85	83 1/2	82 1/2	81	80	78 1/2
10	87	85 1/2	84 1/2	83	81 1/2	80	79	77 1/2
9	86 1/2	85	83 1/2	82	80 1/2	79	77 1/2	76
8	85 1/2	84	82 1/2	81	79 1/2	77 1/2	76	74 1/2
7	84 1/2	83	81	79 1/2	77 1/2	76	74	72
6	83 1/2	81 1/2	79 1/2	77 1/2	75 1/2	73 1/2	71 1/2	69
5	82	79 1/2	77	74 1/2	72 1/2	70	67 1/2	65
4	79 1/2	76 1/2	73 1/2	70 1/2	67 1/2	65	62	58 1/2

Example of combustion efficiency. With a stack temperature of 450°F, and a Co_2 of 8%, the unit would be 81% efficient.

Dwyer Smoke Gage No. 910

Operating Instructions

1. Slip the sampling tube through an opening in the smoke pipe between the draft control and the furnace.
2. Let the burner become fully heated before making a smoke check. Pump the aspirator bulb several times. Insert a piece of sampling paper in the holder and secure it with the sealing plug.
3. Force the smoke sampling through the sampling paper with *thirty full strokes* of the aspirator bulb.
4. Remove the sampling paper from the holder and compare the density or darkness of the smoke deposit with the smoke chart.
5. The smoke chart is a guide to show the maximum amount of smoke which may be given off by each type of burner for clean combustion.
6. It is recommended that air and fuel adjustments be made to increase the CO_2 as high as possible without creating smoke in excess of the limit on the chart for the type of burner being tested. This procedure will give maximum fuel saving efficiency with soot free burning.

Chapter 4

LOW-PRESSURE ATOMIZING BURNERS

Low-pressure atomizing burners found in domestic installations are (1) Williams Oil-O-Matic, (2) Winkler, and (3) General Electric. The last named really should be subdivided with (1) burners used in boiler-burner units and (2) conversion burners.

Williams Oil-O-Matic burner has nine major parts:

1. Shutoff valve, to stop oil flow when burner is not running.
2. Thrift Meter, or metering pump to measure out the desired gph.
3. Pressurotor, or compressor to compress air and mix it with the metered oil.
4. Stabilizer, or float chamber, where mixed oil and air separate.
5. Nozzle, where oil and air are remixed and discharged to the combustion chamber.
6. Combustion-air fan, as on high-pressure burners.
7. Ignition transformer, as on high-pressure burners.
8. Electrodes and spark gap, as on high-pressure burners.
9. Motor, as on high-pressure burners.

The *Thrift Meter* is not relied upon for any suction lift, so fuel should arrive at the burner by gravity. It is stopped at the shutoff valve until the Pressurotor generates over one lb. pressure. This pressure acts on a bellows, which opens a valve and permits oil to flow to the Thrift Meter. This part has a piston in a radial hole in an eccentric rotor. Over one-half revolution the piston moves in one direction to draw in fuel oil; it reverses over the next half revolution and forces the fuel oil on to the Pressurotor. The eccentricity of the rotor is adjustable, thereby changing the gph rate. As the Thrift Meter is a positive-displacement pump, capacity is not related to fuel viscosity.

The *Pressurotor* is a vane-type compressor. The fuel oil from the Thrift Meter is injected into the air being compressed. The main discharge is to the Stabilizer, but a by-pass line from the discharge to the suction side, adjusted by a needle valve, is for regulating the air pressure. Another line connects to the shutoff valve, as has been said.

The *Stabilizer* is a float chamber where the air and fuel discharged from the Pressurotor separate. A line from the upper part of the Stabilizer delivers air to the nozzle. A line inside this line but leading from the lower part of the Stabilizer, and controlled by the float valve, delivers fuel to the nozzle.

The nozzle again mixes air and fuel. It resembles high-pressure nozzles only in having tangential slots that whirl the air and fuel mixture to obtain a hollow cone spray. The orifice is much larger than for high-pressure nozzles—about 1/8 in. as compared with 0.010 in.

The combustion-air fan, ignition transformer, electrodes, spark gap and motor are exactly the same in principle as corresponding parts of high-pressure burners.

To start an empty Oil-O-Matic burner, the service man must first vent the shutoff valve by unscrewing the vent at the top. This should be reapplied when the valve body is full of fuel.

Then the air shutter should be *completely closed* and the burner started. The burner will not fire from the nozzle at once as it takes a short time to build up a supply of fuel in the Stabilizer. The flame will be small at first, but will increase in size. The air shutter should be opened gradually, just fast enough to stop smoking.

There are *two adjustments:* (1) fuel rate, by the Thrift Meter adjusting screw, and (2) air pressure, by the Pressurotor by-pass adjusting screw.

The *Thrift Meter rate (gph)* is increased by screwing in the adjusting screw and decreases by screwing out. Adjustment should be by small increments with at least five minutes allowed to judge effect of each setting, as it takes almost that long for the rate at the nozzle to steady at the new setting.

About one-lb. air pressure is required to open the shutoff valve. The pressure should be somewhat higher than this—1 1/2 to 2 lb. is usual. Pressure should be above the amount at which the fire shows sparks; it should be below an amount that would make combustion noisy. Here again, the burner should be given time to stabilize at any setting before judging it satisfactory or otherwise.

Pressure gage used should be one with a range up to 10 lb., and should be checked for accuracy at a 2-lb. reading. Never should an Oil-O-Matic burner be adjusted without a pressure gage.

An Oil-O-Matic burner that is not functioning may not be generating enough air pressure to open the shutoff valve. If pressure is below 1 1/2 lb., the cause may be too much opening by the bypass valve, clogged muffler, sticking Pressurotor vanes, worn Pressurotor. The muffler to the Pressurotor is the most likely location of the trouble. It can be removed and cleaned.

If air pressure still cannot be built up, a little solvent naphtha introduced through the air inlet may free sticking vanes. If air pressure is still too low and the wear is not excessive, reducing the by-pass of air may do the job. Excessive wear calls for Pressurotor replacement. Assuming that the Pressurotor is functioning satisfactorily, fuel oil may still not be delivered to the nozzle because of other conditions. First of all, there may be a failure of the supply line to the burner. Shut off the line and open the shutoff valve vent cap. If the valve body is full of oil, the indication is that the supply line is working.

With the supply line still shut off and the vent cap removed, start the burner. If the valve stem rises, the indication is that the valve is free and the bellows are intact. If the stem does not rise, it may be stuck. It can be freed by rotating it by a screwdriver.

If the valve stem will not rise, although free, the large bellows may be leaking, or the small bellows may be dirt filled.

If the shut-off valve stem rises yet oil level does not drop in the valve body (with feed line shut off) the Thrift Meter is not handling oil. There may be one of four causes:

1. Thrift Meter piston is sticking.
2. Thrift Meter piston spring is broken.
3. Thrift Meter shaft key is sheared off.
4. Line from shutoff valve to Thrift Meter is clogged.

If the oil level falls in the shutoff valve when burner is operated, the forgoing four possibilities are eliminated. If the burner still does not fire, the trouble then indicated is one of three:

1. Nozzle clogged.
2. Oil tube clogged.
3. Float stuck in lowest position.

There is a bleeder hole in the shutoff valve housing. When the burner is running, a steady flow of *fuel oil* from this hold indicates a leaky small bellows; a flow of *fuel and air* indicates a leaking large bellows.

If the shutoff valve does not seat tightly, the Stabilizer may fill up in off periods. This causes an abnormal fire at starting. There will be flame roar and odors. The cure is to repair the valve. Opportunity should be taken to inspect both bellows while working on the valve.

The *Winkler burner* operates on a slightly different principle. There is no intermediate separation and subsequent remixing of fuel and air. Both are fed to the nozzle in a single tube.

A shutoff valve, opened by fuel aerator (compressor) air pressure admits fuel from the supply line via a strainer to the *Fuel Meter* (Metering Pump) body. The Fuel Meter is of duplex plunger type, driven by a shaft connected to the motor shaft by a worm gear. The plungers not only reciprocate, they also turn with each stroke so as to uncover intake and discharge ports at the proper times. This pump should not be relied on for any suction lift. If the tank is not located so that the fuel oil flows to the Fuel Meter by gravity, a supply pump can be added to the burner. An auxiliary Wall pump of convential type could be used instead, but would probably be more costly.

The Fuel Meter has a fixed capacity, stamped on the body. As the pump is positive in displacement, the rate is independent of viscosity. To change firing rate, the plungers and plunger housing must be changed. The oil lines should be uncoupled and six screws removed. This drops the lower housing. Plunger cotter-pin removal permits removing plungers. Five capacities are available: 0.50, 0.75, 1.00, 1.25, and 1.50 gph.

Fuel discharged from the Fuel Meter is sent to *Fuel Aerator* or compressor, where it is added to the air being compressed. This aerator is an eccentric cam with a sliding divider between inlet and discharge. Air pressure produced is about 3 1/2 pounds.

Fuel and air from the aerator goes to a *Percolating Chamber,* which is a small chamber designed to give the mixture a sudden 180 degree turn. This agitation has a homogenizing effect.

The percolating chamber is connected to a larger space—the *air cushioning chamber*—that smooths out the pulsations of the Fuel Aerator and supplies pressure to the line that opens the shutoff valve.

From the percolating chamber, fuel and air go down the line to the nozzle. At the latter, a venturi speeds up velocity and the fuel and air are put through several 90-degree turns. Finally, the mixture is discharged through 4 tangential channels and an orifice, giving a hollow-cone spray. The orifice is 5/32 inches in diameter.

Combustion air is furnished by a convential *fan,* regulated by a shutter called the *air regulator.* The air head, however, is somewhat unique. The main head, called the *Flame Controller,* is open at the back. The electrode holder closes the center of the back entrance and has spiral vanes around its circumference. To get inside the Flame Controller, air must pass through these vanes. It then goes forward, around the nozzle, and into the combustion chamber.

The outer circumference of the Flame Controller also has vanes, spiraled in the same direction. Air delivered between the blast tube and the Flame Controller must pass through these vanes and then inward over the nozzle. At the front, the Flame Controller has an inner cup and some air passing between the Flame Controller and the blast tube can enter the annular space between controller and inner cup through a series of radial holes in the former. This air is not whirled and enters the chamber in between the two whirled air deliveries.

The Flame Controller position fore-and-aft can be adjusted. Moving it forward lengthens and straightens the fire. Moving it back, shortens and widens it.

Only adjustments possible on a Winkler burner are amount of combustion air by means of the fan regulator and position of the Flame Controller. A flat spot on the top of the housing is for the purpose of *leveling* the burner. It should be leveled both *fore-and-aft* and *athwart*. This is quite important. The end of the blast tube (blast-tube hood) should be 1/8 to 1/4 inches *back* from the inside of the combustion chamber.

Two flat surfaces on the back end of the nozzle should be vertical. If they are not the fire may be heavier on one side. If after starting the burner, one finds the fire one-sided, the nozzle should be rotated so the top is turned very slightly to the lean side of the fire.

If a Winkler burner is started up empty or partially empty of fuel oil, it should be bled by backing out a bleed screw near the top of the fuel meter. After a small amount of solid oil has escaped, the screw is again tightened.

The strainer can be cleaned by removing the drain plug below it and allowing a quart or two of fuel oil to escape through it. After the plug is replaced, the burner should be bled.

The *General Electric* burner has separate oil and air feeds from a separating sump and in that respect is similar to the Oil-O-Matic burner. It meters fuel by means of an orifice in the nozzle, and thereby differs from both Oil-O-Matic and Winkler.

The boiler-burner unit is made in seven sizes:

Model	LA20	LA22	LA32	LA34	LA42	LA44	LA54
GPH	0.70	0.94	1.33	1.91	2.65	3.37	4.25
EDR sq. ft. (1 pipe stream)	195	293	410	586	806	1,025	1,320
Oil Pump gph max.	6	6	6	6	6	9	9
Fan Capacity, cfm	100	100	100	100	100	225	225
Fan pres. in.	1.0	1.0	1.0	1.0	1.0	1.1	1.1

Fuel oil enters the unit via a *screen valve,* which is a strainer and antisyphon valve. This delivers oil to a *motor compressor* which also compresses air. The motor compressor discharges to a *sump* where air and fuel separate. The intention is to have a fairly constant level of fuel oil in the sump under a constant air pressure. The first aim is secured by means of a float, which rises if too much oil accumulates and lifts a valve that opens the fuel suction line to the air space of the sump, thereby decreasing the intake of fuel to the motor compressor. A falling oil level causes the valve to close, thus increasing oil flow. The float will stabilize somewhere between valve full open and closed.

The second air—constant air pressure—is achieved by a bellow-controlled needle valve (*pressure-regulating valve*), which, when open, allows some sump air to escape to the motor-compressor air inlet. This valve is adjustable, but the adjustment should usually be for 15 lb. in combination units, although it may be as low as 13 lb. at a slight decrease in gph capacity. It is this pressure and the nozzle orifice that does the metering.

Oil from the bottom of the sump and air from the top are led by two lines to the *nozzle*. Admission of oil is controlled by a *solenoid fuel valve* timed to open after air flow begins. Air is introduced after oil is metered through air *orifice* and at right angles. The mixture then passes through two orifices, the latter leading to the combustion chamber.

In the burner used for boiler-burner and furnace-burner units, there is no whirl introduced, so the nozzle produces a *solid* cone. Combustion air is supplied by a centrifugal *fan*, with amount of delivery controlled by a *regulating disc*. On combination units, part of the fan delivery is directed down the burner tube past the nozzle. The remainder is piped to the bottom of the boiler or furnace and directed up against the flame through an *air distributor*. The effect of this is to turn the cone tips up to the top sides of the steel combustion chamber.

Ignition transformer is conventional. General Electric states that persistent *radio interference* may be eliminated in some cases by using an *interference filter*, which is 2 1/2-mf condensers in series with the midpoint grounded. Such a filter may be put across the incoming hot and ground lines, or across the primary winding of the ignition transformer, or across thermostat terminals, depending upon the cause of the interference. In each case, the midpont of the filter should be grounded. It is also suggested that circulating-fan motor frames and fan housings that are mounted on rubber be grounded.

The suction line to a General Electric burner should be brought in overhead. It should go vertically up from the tank, then slope gradually up to a point over the burner, then go vertically down to the burner. The highest point should be at the top of the vertical line down to the burner. If this layout is deviated from, fuel oil from the lateral run may flood the sump, during off periods.

A failure to maintain sump pressure may be caused by

1. restrictions in air inlet
2. leaking sump gasket
3. leaks in air line or burner head
4. defective pressure regulator
5. defective pump

Suction-line leaks may be located if a vacuum gauge is connected. If there is a leak, the vacuum will probably be less than 19 inches. Beginning at the tank, stop off the line where possible. If at any location, the vacuum improves, the leak is on the last section to the tank side. Leaking screen-valve diaphragm may be tested by stopping the hole in its face. If the vacuum then goes up, the diaphragm leaks. The screen valve may have a leaking diaphragm, may open too hard, not close tightly, or have a defective casting.

Former models had two-piece *porcelains or electrodes*, with the latter bare between pieces. Sometimes the bare electrode shorted to the burner tube. If this has been happening, old porcelains should be replaced with newer one-piece porcelains. The so-called ignition barrier should also be changed as the old design will not fit new porcelains. The new ignition barriers are of fiber and eliminate arcing from electrode connectors to the air tube.

Timken manufactured both vertical-rotary and gun-type burners. Various boiler-burner, furnace-burner and water-heating units are also manufactured. The only burner that will be described is the vertical-rotary Timken line.

Model	Oil Rate lb/hr min/max	Motor hp	No. of Ignitors	Power Aver. Watts	EDR Steam
J	2/7	1/100	1	63	400
F & OCA-10	5/10	1/40	1	58	575
H & OCA-18	8/18	1/30	2	75	1,035
C & OCC-50	10/50	1/20	2	88	2,600

	Round Boiler/Furnace		Rectangular Boiler/Furnace	
	Min Diam.	Max Diam.	Min Width	Max Width
J	(furnished in units only)			
F	15″	22″	15″	20″
H	17″	27″	17″	24″
C	17″	40″	17″	40″

All models are furnished standard with motors for 60 cycle, 110 volt, single phase current, 1,750 rpm. Motors below 1/20 hp are of shaded-pole type. Motors of 1/20 hp are of split-phase type. Models F, H and C are for intermittent ignition; Model J is for constant ignition.

Fundamental Principles

Although there are minor differences in construction as shown by the various models, essentially any model consists of:

1. a constant-level valve (Timken "L-8" valve) to provide a supply of fuel oil at a fixed level slightly higher than the level desired in the burner to supply oil at a constant rate.
2. an "oil-valve," which is a combination of solenoid shutoff valves and a metering valve, to stop or open up fuel oil flow as desired and meter the quantity.
3. the burner proper, which is
   a. a stationary outer assembly consisting of supports, housing, air shutter, and motor stator, and
   b. an inner rotating assembly, consisting of motor rotor, oil-delivery tubes and fan.
4. A flame rim, namely a circular or rounded-corner rectangular band of steel inside the boiler of furnace walls, to intercept and provide for mixing of air with the oil stream thrown out by (3)
5. Ignition system, consisting of ignition transformer, electrodes and wiring.

Constant-level L-8 valve is conventional, with a float-operated valve to hold a constant oil level in the float chamber, which discharges to the supply line to the burner. Figures 4.1 and 4.2 show how the valve should be connected to an inside storage tank. Figure 4.3 shows relations between levels of tank outlet, constant-level valve inlet and oil-discharge level in burner.

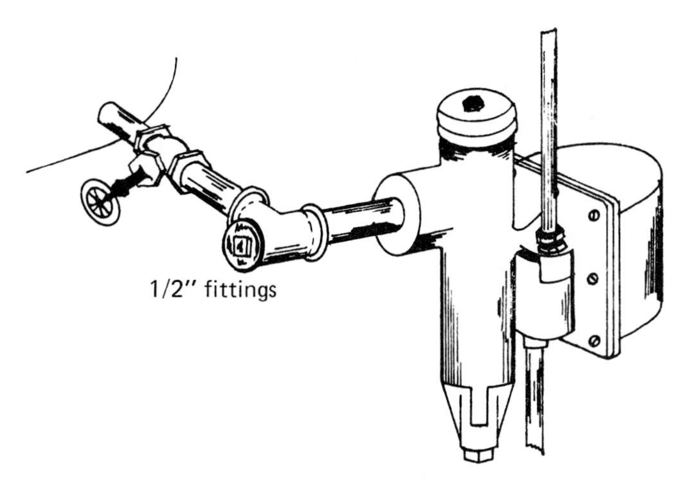

Figure 4.1. L-8 valve piping, fuel rates up to 22 lb/hr.

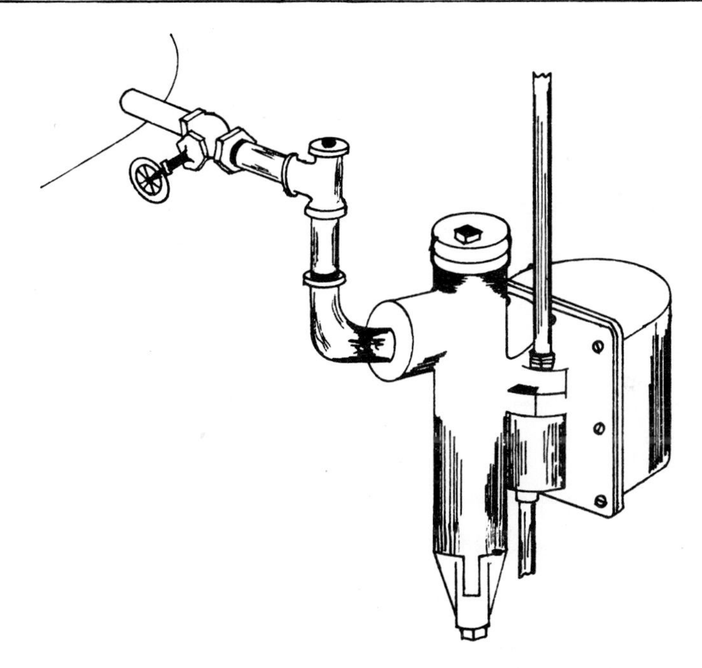

Figure 4.2. L-8 value piping, fuel rate 22 lb/hr or higher.

Oil Rate	Oil Line			
	17′-3/8″ O.D. Tubing and Timken Oil Value		17′-1/2″ Iron Pipe, 4 Ells and Timken Oil Valve	
	R	S	R	S
5	0	3″	0	—
10	0	4 1/2″	0	—
15	0	6″	0	4 1/2″
22	0	8 1/2″	0	6″
30	4″	12 1/2″	4″	9″
35	4″	15 1/2″	4″	10 1/2″
42	4″	19 1/2″	4″	13 1/2″
50	—	—	6″	17 1/2″

Oil line—drop—directly to the floor from L-8 outlet, and pitch up continuously from oil valve outlet to the burner.

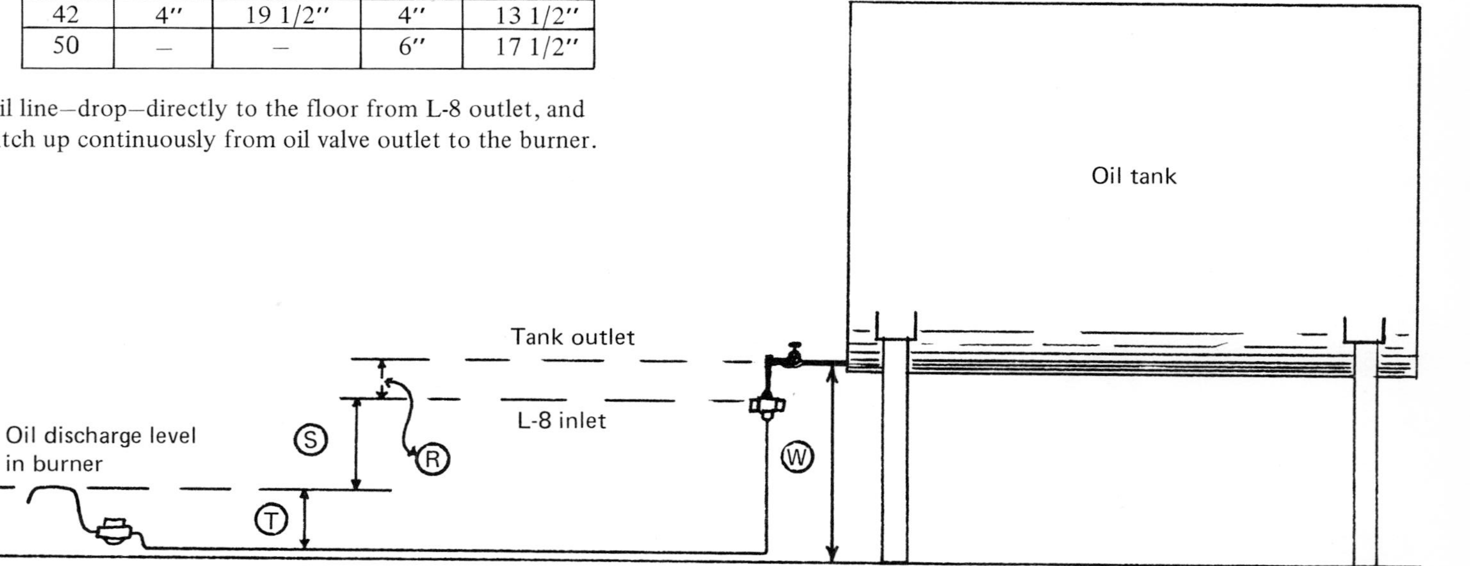

Figure 4.3.

The L-8 constant-level valve has a vent connection that should be piped independently, terminating above the tank. It should not be connected into the tank vent.

On steel-hearth units (Models 30E, 40A and 50B water heaters; BAR 3 and BER oil-boilers; FER-65 and FCR-100 oil furnaces) the L-8 valve is not used, as there is what is called a "gravity control valve" in the base of the unit.

This is a constant-level valve with a trip-bucket safety. Failure of the float valve to seat, or oil accumulation on the hearth because of lack of ignition, will result in action by the trip-bucket to shut off oil flow.

Oil Valve consists of two solenoid-actuated shutoff valves in series with a hand-operated needle-type metering valve to regulate amount of flow. Construction is shown by Figure 4.4. The solenoids are wired in parallel with the burner motor so that they open their valves as the motor starts and close when the motor stops.

Although the metering valve is adjustable, the amount of permissible adjustment once the burner is installed for a hearth of specific size and shape is small. Within capacity limits of the burner, efficient firing rate is fixed rather closely by the hearth. Variations in heating requirements must be taken care of by variation in length of running time, as in this case of gun burners.

In Figure 4.3, one of the significant levels is that of the oil delivery in the burner. This level is not that of the outside oil connection on the burner, but is 1 in. above the outside connection on Model F and OCA-10, 2 in. above on Model H and OCA-18 and 6 in. above on Model C & OCC-50. These differences should be allowed for laying out the supply system and locating constant level valve.

Note from Figure 4.3 that the supply pipe should drop vertically from the L-8 valve with not less that 1/2 in. pipe or copper tubing, be carried horizontally to the oil valve, and then pitch continuously up to the burner. This is to avoid formation of air pockets.

On the installation of the burner the supply line should be vented before connection to the burner. With the metering valve wide open, the solenoid valve is energized and the oil allowed to flow until free of bubbles. If venting should become necessary after installation (tank allowed to run empty, for example) it is preferable to break the line at the oil-valve strainer and not disturb the setting of the metering valve.

Burner in the restricted sense is the rotating and stationary assembly as outlined under (3) in the foregoing.

Cross section of Model F is shown in Figure 4.5, of Model H in Figure 4.6 and of Model C in Figure 4.7a. The exact construction differs, but the principle is the same: namely the fuel oil is delivered inside a rotating member, given thereby a rotating motion and subjected to centrifugal force. Because of this force, the oil climbs up to find a greater diameter, eventually enters the oil tubes and is flung out against the flame rim, much as water leaves a rotating lawn sprinkler.

On Models F and H the lower rotating chamber is a cone tapering out and up and the relatively long tubes are connected to the top of this chamber. On Model C, the tubes are much shorter and the rotating cone is quite short in elevation but of greater capacity. On Models F and H the entire rotating assembly, including the motor rotor, can be lifted out. On Model C the assembly cannot be lifted out but must be taken apart to be removed.

Lubrication is not required on Models F and H as they are self-lubricated by the fuel oil being handled. Model C has two oil cups that should be filled at intervals of 3 to 6 months with SAE 10 lubricating oil.

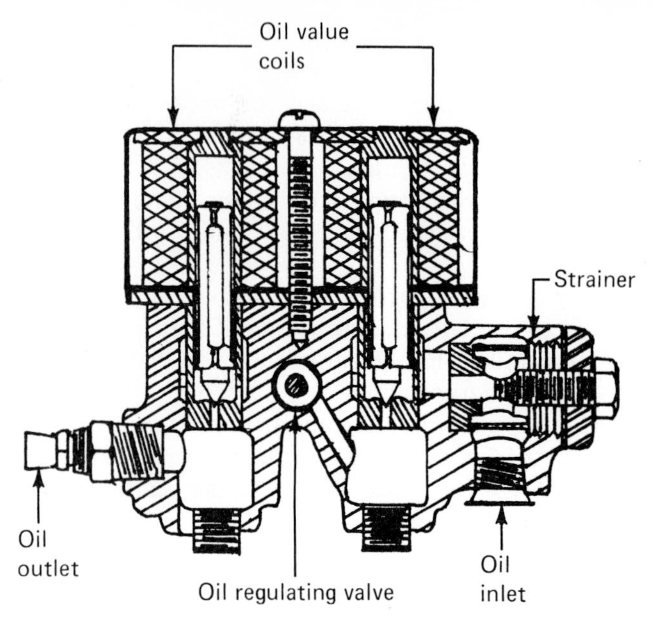

Figure 4.4. Oil valve.

Fan is of centrifugal type (see Fig. 4.8) with straight blades slanted slightly back from direction of rotation at the outer ends. Purpose of the fan is primarily to support the oil delivery streams and secondarily to furnish combustion air. Fan should deliver (1) proper amount of air and (2) at proper velocity. Amount of air is governed by a conventional air band in the stationary part of the burner. Velocity of delivery is a matter of fan diameter, as fan speed is fixed by motor speed at 1,750 rpm.

There should be clearance between fan body and oil distributing tubes. The tubes should also deliver oil to the flame rim at the same level and over a narrow band. If the delivery of the tubes does not coincide, one of them should be slightly bent until there is coincidence.

If the oil delivery is above the top of the outer flame rim or below top of the inner one satisfactory ignition will not be obtained. To provide prompt ignition the oil must impinge on the flame rim at a point just above the ignition spark. Level of delivery can be changed on Model C by adjusting burner motor, on Model F by shifting burner body, on Model H by shifting motor stator.

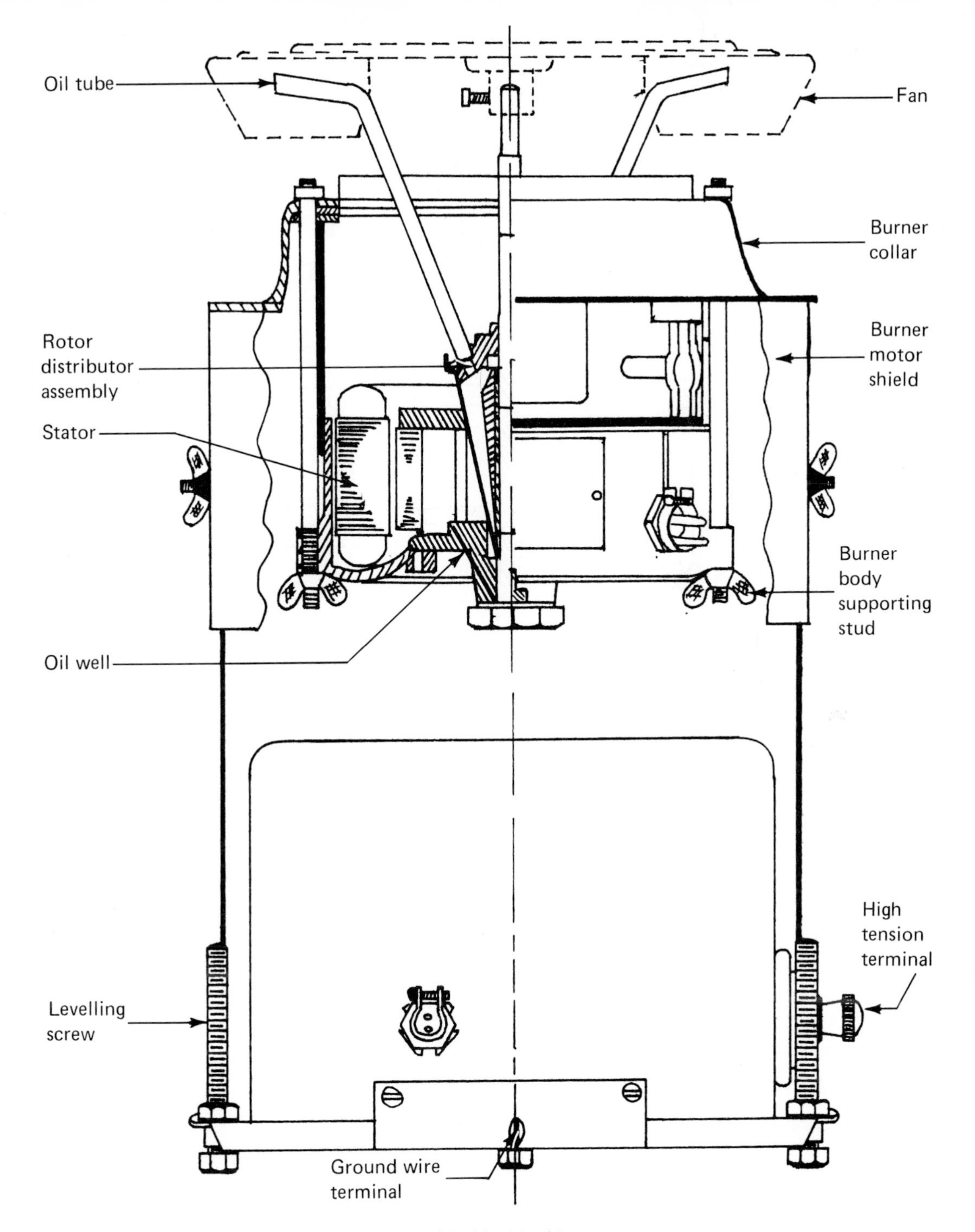

Figure 4.5. Model of burner.

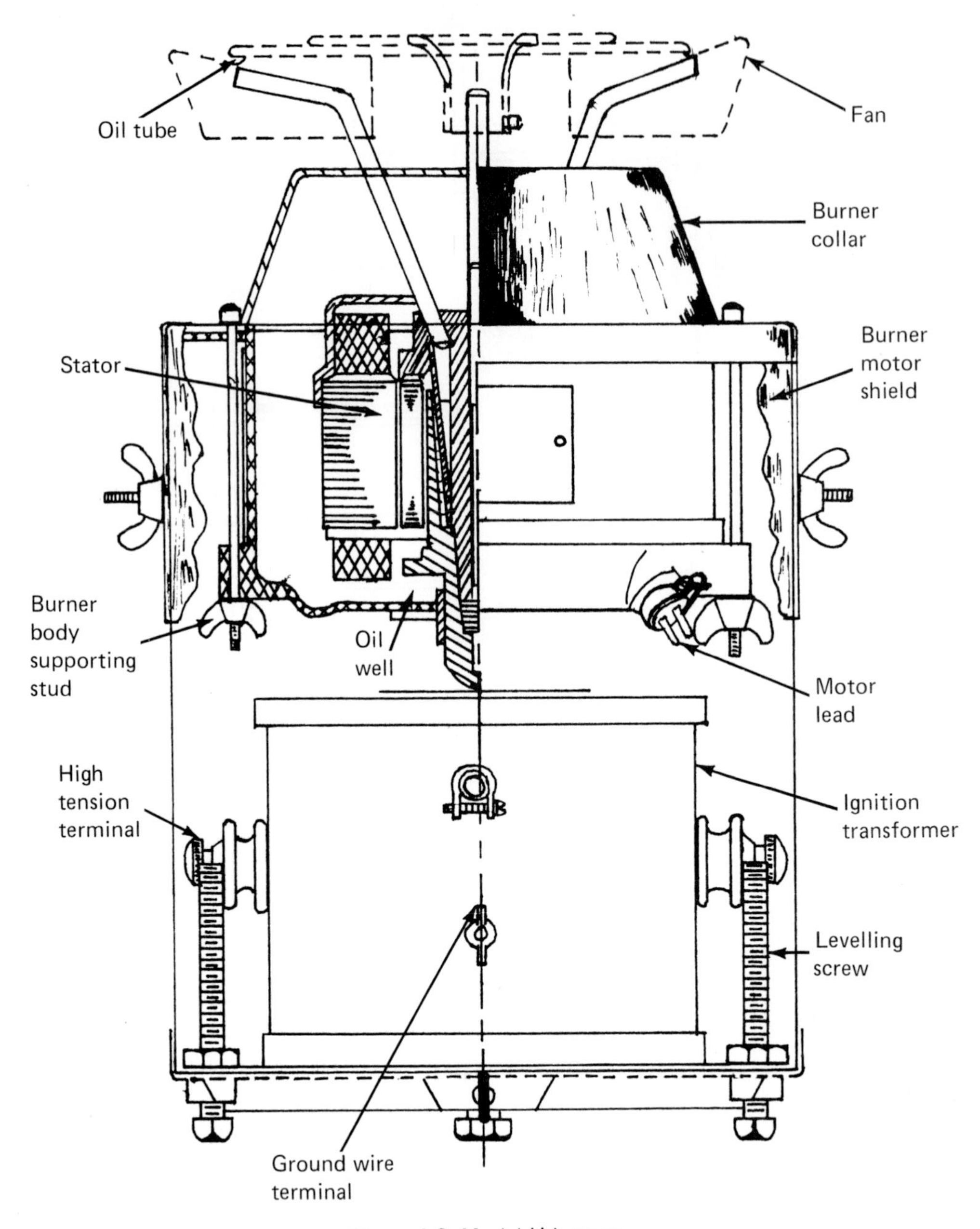

Figure 4.6. Model H burner.

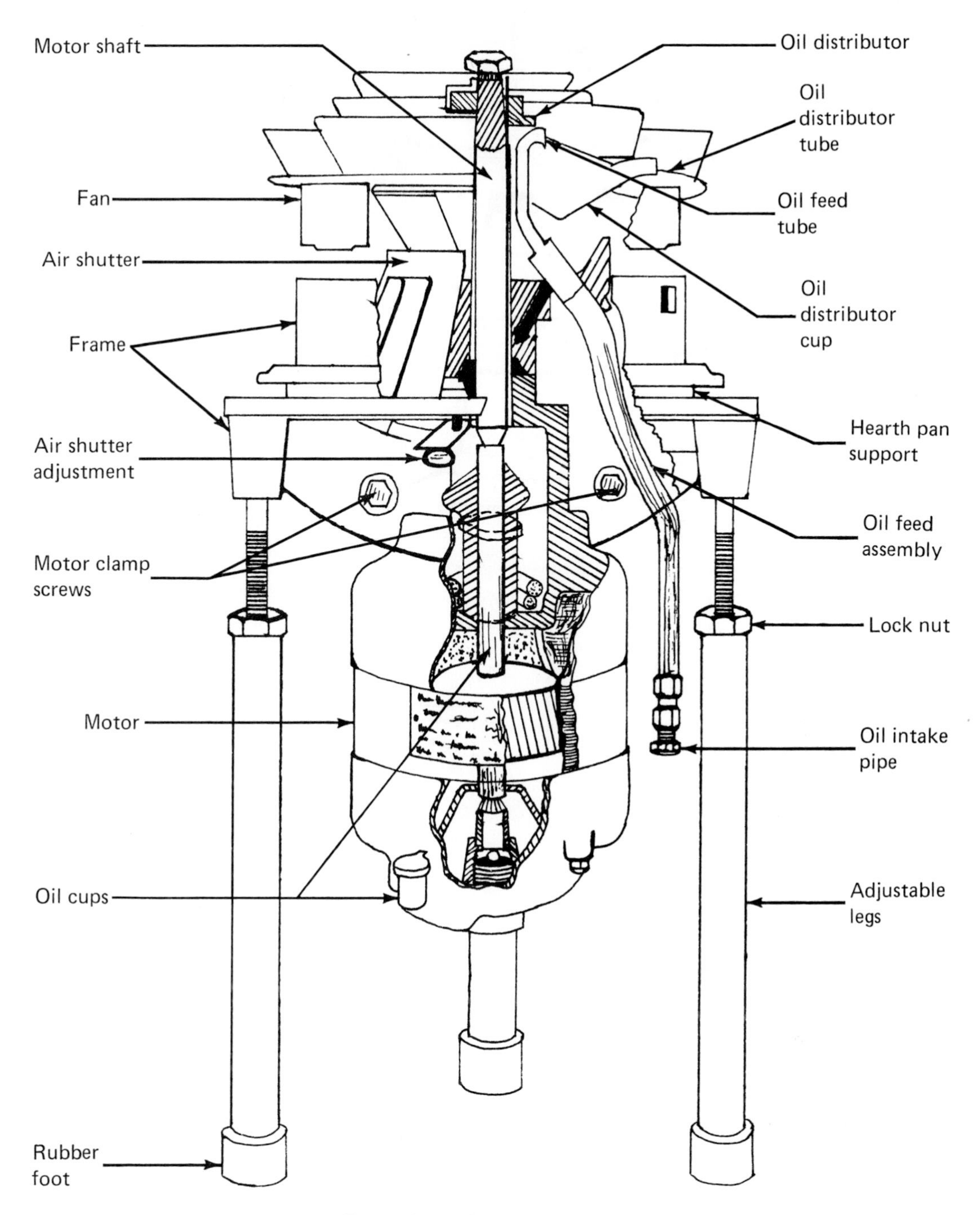

Figure 4.7a. Model C burner.

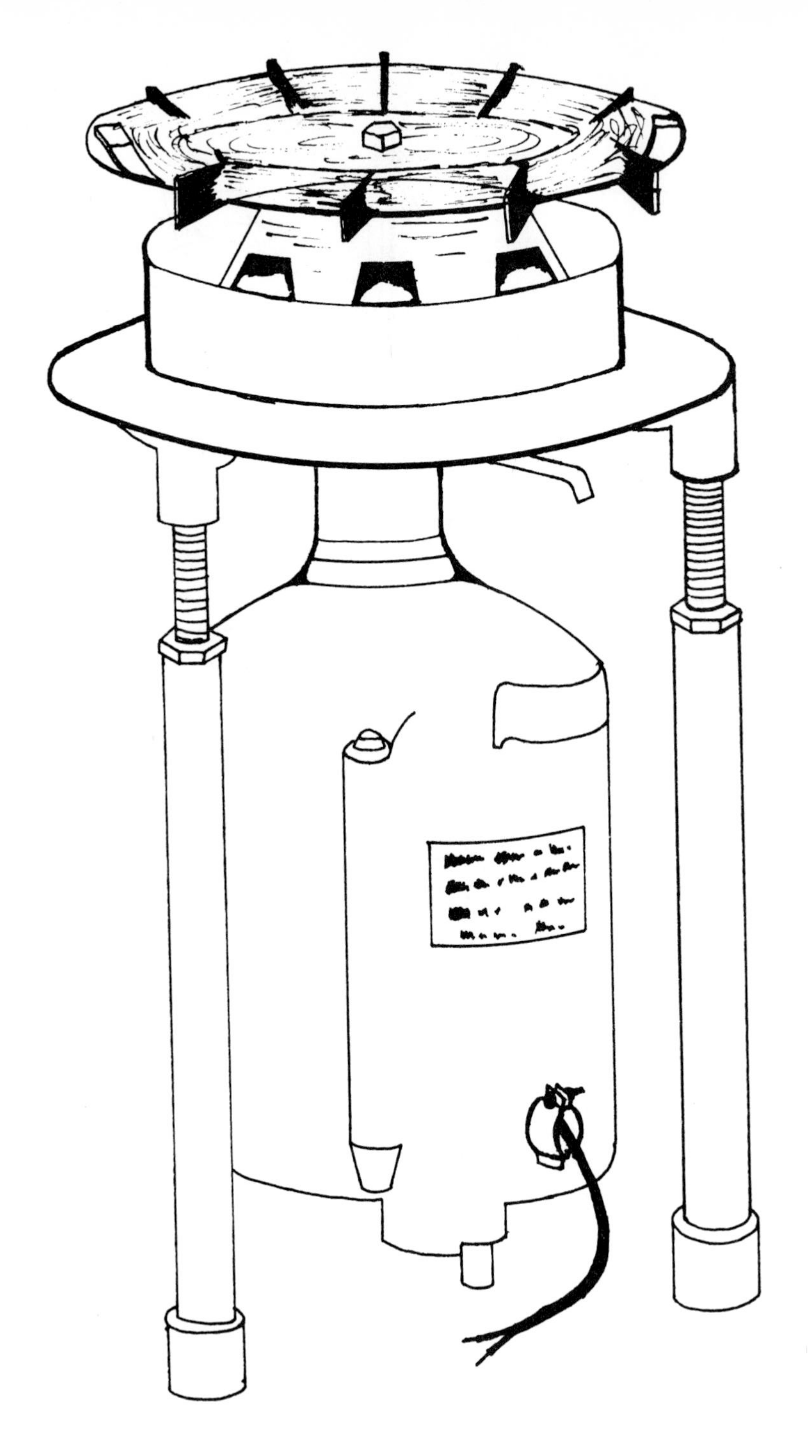

Figure 4.7b. Model C burner.

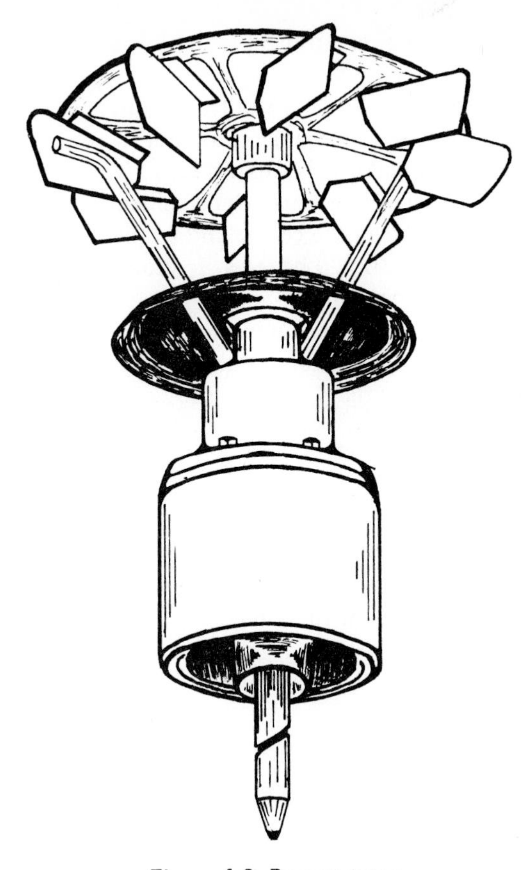

Figure 4.8. Burner rotor.

Each model takes a variety of fan sizes depending upon oil rate:

Model	Capacity 1b/hr	Fan Size Number
F	5/8	1-68
	9/10	2-68
OCA-10	5/8	3-68
	9/10	4-68
H and	8/14	1-75
OCA-18	15/18	2-75
C and	10/12 1/2	1-60
OCC-50	12/18	2-60
	16/30	3-60
	30/40	4-60
	40/50	5-60

All models have leveling screw or adjustable legs. It is very important that burners be installed level.

Sometimes it will be found that the fan installed produces too high an air velocity for the firing rate. The permanent remedy is to change fans, but a temporary repair is to bend blade tips back slightly, being careful to make all bends alike.

Hearth is constructed of special refractory-type cement on a steel hearth plate, that fits around the base of the burner collar in the center and to the boiler or furnace wall at the outside. The hearth on current models should be level. It should be deep enough that the top surface is flush with the top of the burner collar.

The Flame Rim is a chromium steel through Models F, H, OCA-10 and OCA-18; the inner upright section about 1 in. high and the outer about 2 in. high, the two sections joined by a bottom flat section about 1 in. wide (see fig. 4.9). The oil streams from the burner should clear the inner wall but strike the outer one.

The Flame Rim is usually circular for round or square boilers. For rectangular boilers, the rim is oblong shaped with flat sides and rounded ends, as in Figure 4.10.

For Model C and OCC-50 burners the rim construction is slightly different, as shown in Figure 4.11. Note that instead of a continuous rim, there are sections.

Steel grids or grills are mounted on the outer band of the flame rim. Number of grills recommended is for 1 1/4 to 2 per lb. of fuel oil per hr., the higher ratio applying to the lower firing rates. Grills are designed to permit of various positions on the flame rim and final arrangement should be selected only after combustion test. The positions are shown in Figure 4.12.

Hearth Wall Ring is necessary in square or rectangular boilers. It is made of chromium steel and 4 to 6 in. in height, and is located 1 1/2 in. behind the outer flange of the flame rim, as shown in Figure 4.10 and 4.11.

Ignition Transformer is conventional with exception that models having one igniter only use a 8,500-volt transformer, but models using 2 igniters use a 14,000-volt transformer with a mid-point ground.

Each high-tension terminal (of a 2-ignitor burner) is connected to an ignitor, which is cased in an insulator, which in turn extends through the hearth and sets flush with the top of the hearth (see fig. 4.13). The electrodes should be turned in the direction of burner rotation and in this position the tip should be 1/16 in. above the level of the Inner flame rim band and within 3/16 to 1/4 in. of the outer band.

The flame rim is connected to the ground terminal of the transformer.

Ignition period should be 45 to 75 sec.

Ignitors should always be covered when the boiler is being cleaned to avoid scale or carbon lodging in them. It is also advisable to cover the burner head.

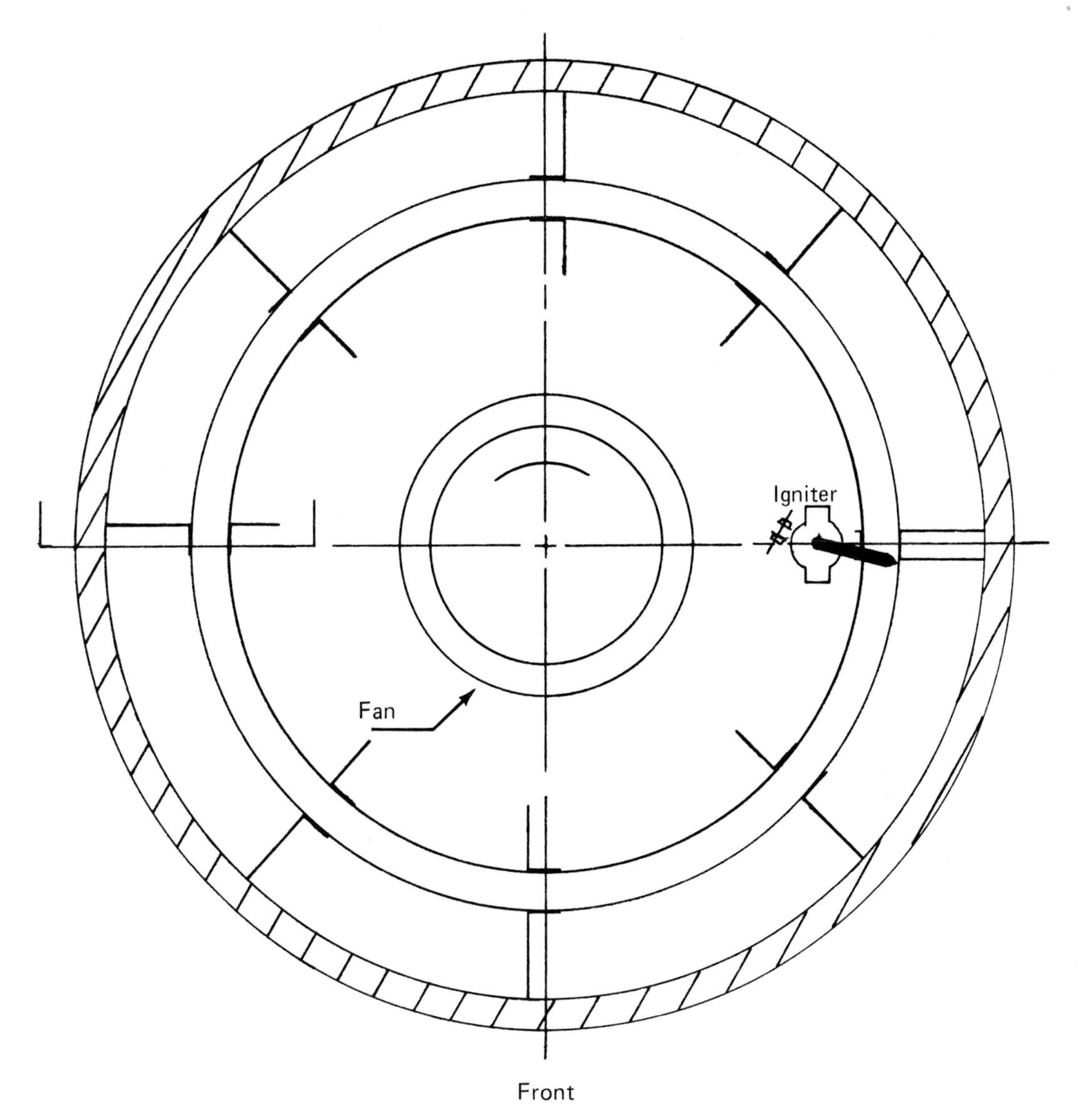

Figure 4.9. FA hearth design.

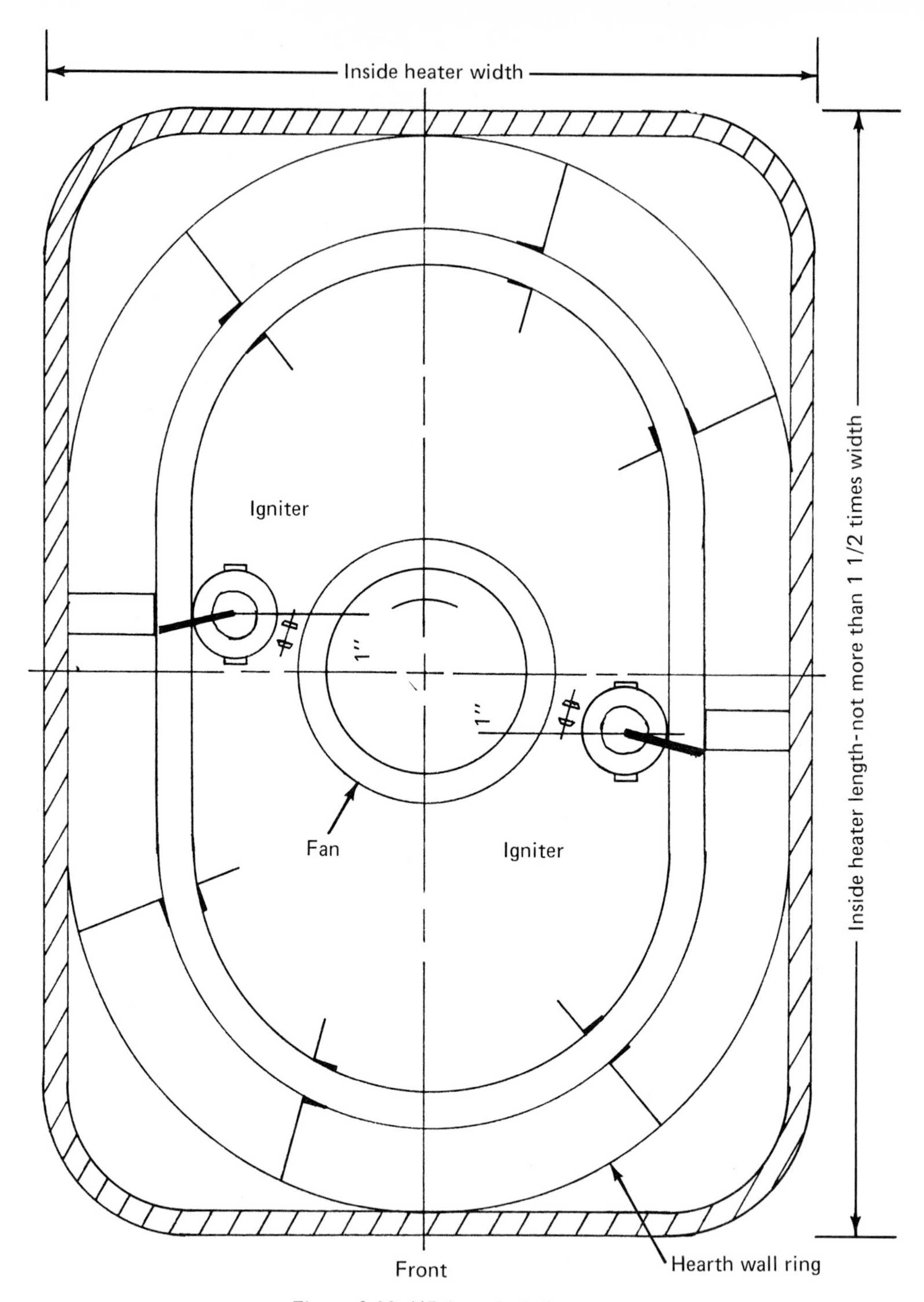

Figure 4.10. HD hearth design.

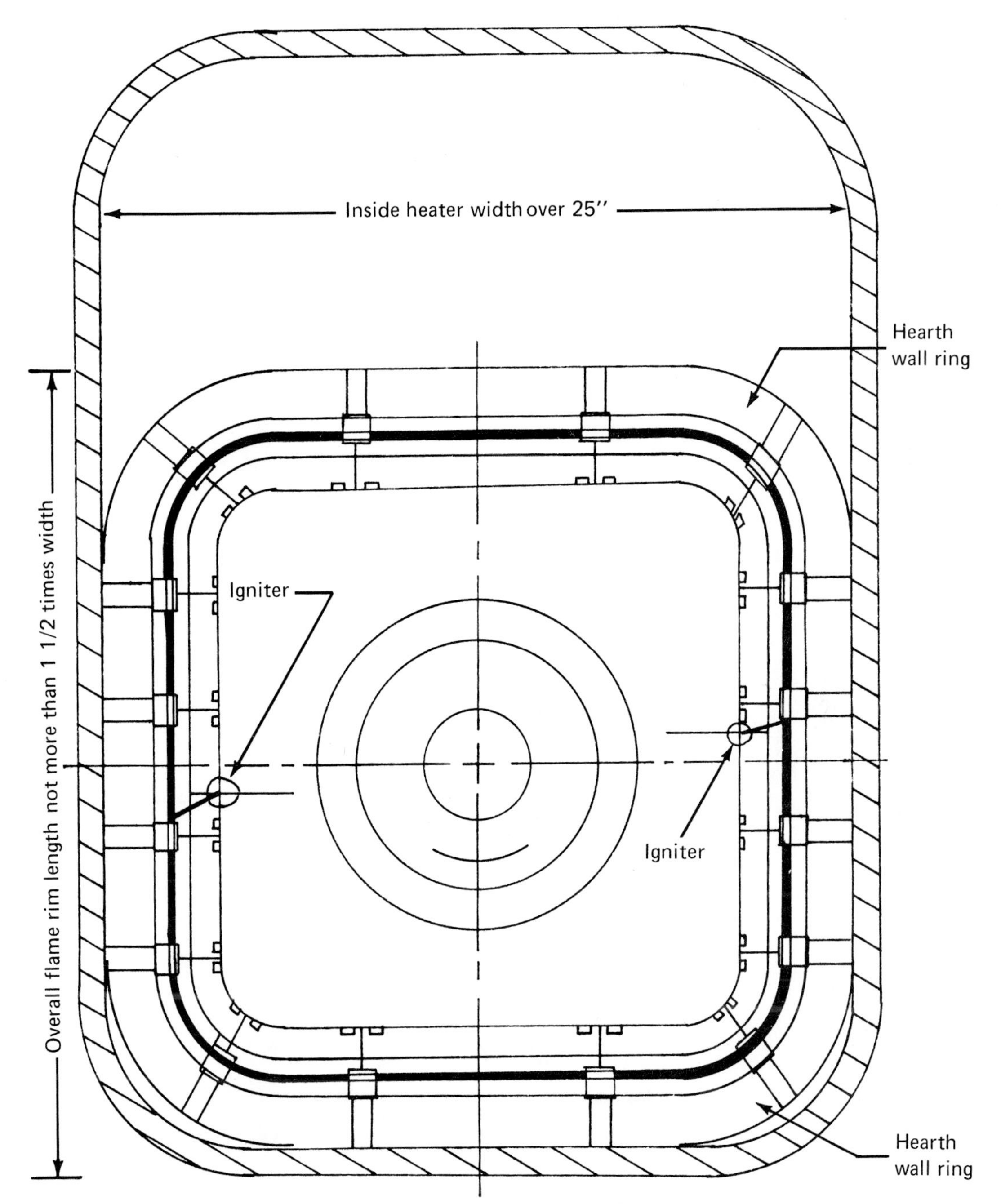

Figure 4.11. CD hearth design.

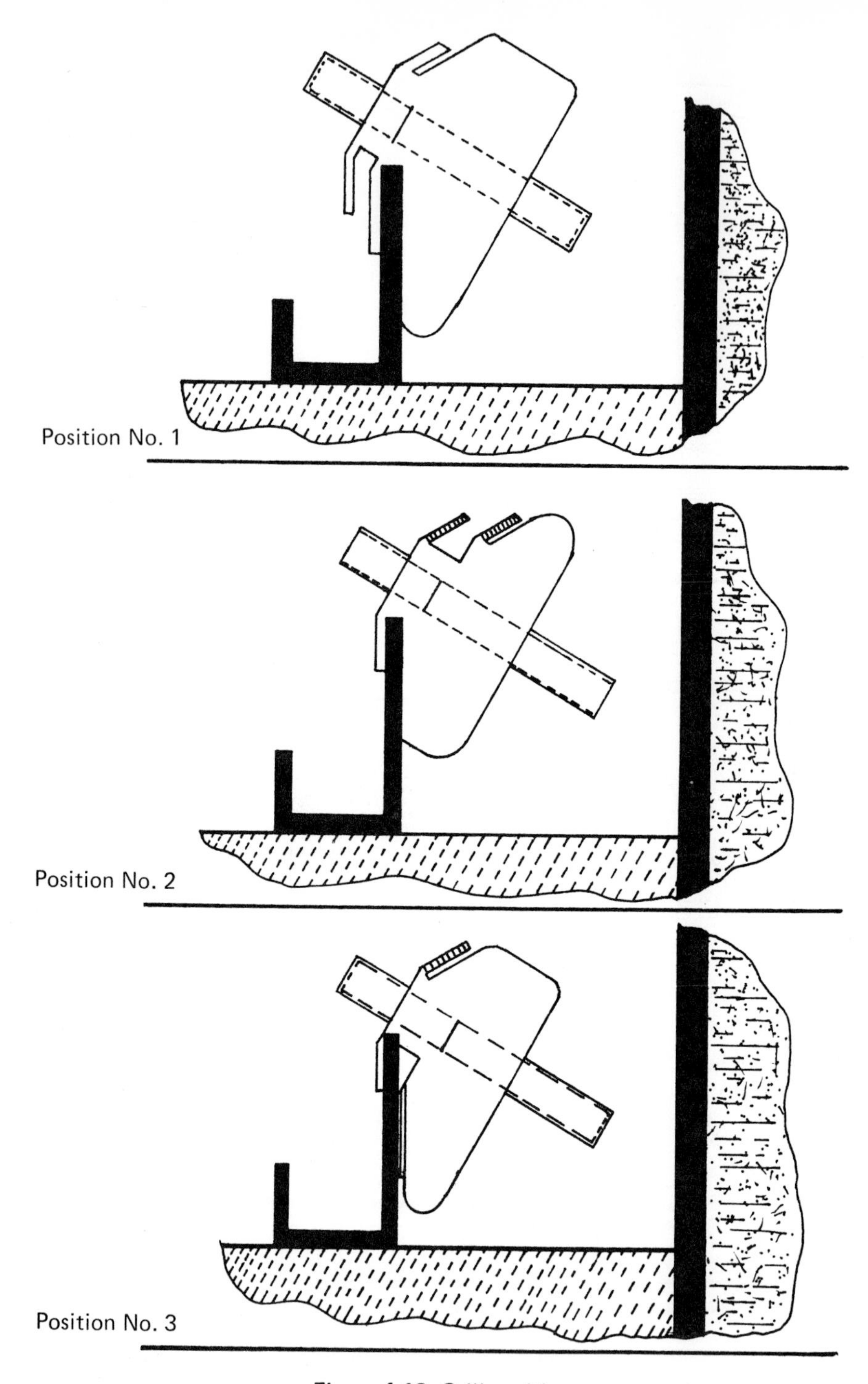

Figure 4.12. Grill positions.

Adjustments

Difficulties encountered with boilers, furnaces, heating systems, controls, etc., are the same for Timken as for any type of burner. There are some adjustments (and difficulties met if much adjustments are not properly made) peculiar to Timken type of burner, however, namely:

Flame

a. Good: blue at base, blue and yellow in body, clean orange or luminous tips, burning quietly and steadily from the grills.
b. Unstable dirty blue flame with dirty streaks with tendency to snuff out on cold starts: air velocity may be adequate but air supply is insufficient. Increase air shutter opening.
c. Flame burns noisily in groove of flame rim: air velocity may be satisfactory but air supply is too great. Reduce air shutter opening.
d. Lazy flame that floats away from grills of air supply is reduced: fan does not provide sufficient velocity. Replace with large fan.
e. Short, choppy flame that becomes gassy and malodorous if air supply is reduced: Fan gives velocity too high. Replace with smaller fan, or bend blade tips back slightly.
f. Rapid pulsations: May be caused by excessive fan velocity, excessive air supply or excessive draft. If pulsations persist after all three are corrected, admission of secondary air should be arranged by making 4 3/4-in. holes through hearth at 90 degrees and near burner collar.
g. Slow Pulsations: May be caused by insufficient air supply, insufficient fan velocity or over-firing.
h. Fluctuating Fire: may be caused by slow ignition, dirty oil distributor tubes, dirt inside oil pickup cone in burner, air or dirt in oil supply line, constant-level valve not holding oil level. This last trouble might be more acute when tank is full.

Puff Backs: May be caused by fan velocity too high, draft insuffficient, downdrafts, slow ignition.

Burner Noise: May be caused by loose fan, unbalanced fan, fan scraping on hearth, motor bearings loose, burner leg or legs loose (see "Trouble Shooting" for causes of noise common to all burners).

Carbon Deposities: May be result of incorrect adjustments as for any type burner, but here look also for air leaks, particularly if extensive enough so that burner air shutter must be too little open for good combustion readings.

Fire Snuffing Out: Excessive draft. Fan velocity too high.

Draft: Should be from -0.02 in. to -0.04 in. over the fire.

No Ignition (or poor ignition, to the extent applicable) may be result of:

a. Carbonized electrodes, probably because there is air leakage around them, or because oil delivery is not correctly placed.
b. Electrodes shorted by boiler scale.
c. Broken electrode insulator.
d. Defective wiring, shorts, grounds.
e. Defective transformer or controls.

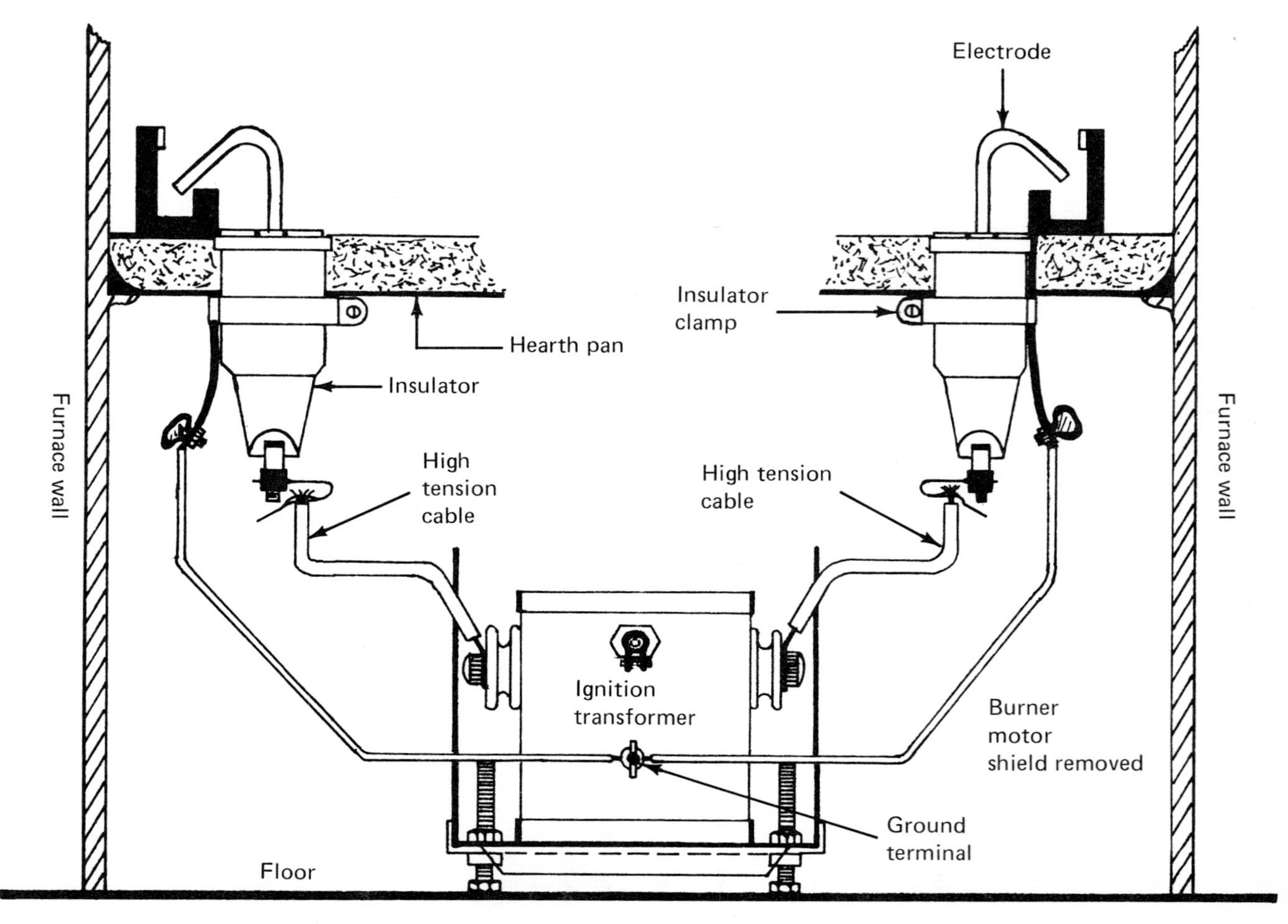

Figure 4.13. Ignition system on model H and C burner.

f. Incorrect spark gap (if from burned tip, should not be spaced by turning electrode).
g. Hole burned in flame rim at spark gap. (Flame rim repair clips are available.)
h. Hearth surface not uniform or not leveled with burner.
i. Excessive fan delivery velocity.
j. Excessive draft.
k. Fuel firing rate too low for hearth size.
l. Restrictions in fuel system, including oil distributing tubes.
m. Hearth wall too low or too far away from flame rim.

Chapter 5

OIL WIRING DIAGRAMS

The following electrical diagrams are typical of oil fired residential equipment.

Figure 5.1a

Oil fired forced air furnace, consisting of

1. Single speed blower motor
2. Combination Fan and Limit Control
3. Heat sensing safety control, constant ignition
4. Heating Thermostat
5. Oil burner motor
6. Oil burner ignition transformer

Figure 5.1b

Figure 5.1b is the schematic of Figure 5.1a

Figure 5.2a

Oil fired forced air furnace, consisting of

1. Heating Thermostat
2. Heat sensing safety control, Intermittent Ignition
3. Single speed blower motor
4. Combination Fan and Limit Control
5. Oil burner motor
6. Oil burner Ignition Transformer
7. Electronic Air Cleaner
8. Humidistat
9. 115 Volt Humidifier

Figure 5.2b

Figure 5.2b is the schematic of Figure 5.2a.

Figure 5.3a

Oil fired forced air furnace, prepped for air conditioning consisting of

1. Fan center
2. Heating-Cooling Thermostat

3. Combination Fan and Limit Control
4. Heat sensing safety control, Intermittent Ignition
5. Oil burner motor
6. Oil burner Ignition transformer
7. Electronic Air Cleaner
8. Humidistat
9. 115 Volt Humidifier
10. Single speed blower motor

Figure 5.3b

Figure 5.3b is the schematic of figure 5.3a.

Figure 5.4a

Oil fired forced air furnace, prepped for air conditioning consisting of

1. Multiple speed blower
2. Fan Center
3. Heating-Cooling Thermostat
4. Combination Fan and Limit Control
5. Heat sensing safety control, intermittent Ignition
6. Oil burner motor
7. Oil Ignition Transformer
8. Electronic Air Cleaner
9. Humidistat
10. 115 Volt Humidifier

Figure 5.4b

Figure 5.4b is the schematic of Figure 5.4a.

Figure 5.5a

Oil fired forced air furnace, prepped for air conditioning consisting of

1. Multiple speed blower
2. Light sensing safety control
3. Fan Center
4. Heating-Cooling Thermostat
5. Combination Fan and Limit Control
6. Oil burner Motor
7. Oil Ignition Transformer
8. Electronic Air Cleaner
9. Humidistat
10. 24 Volt Humidifier

Figure 5.5b

Figure 5.5b is the schematic of Figure 5.5a.

Note, on most light sensing safety controls, the wiring color code is as follows. Black—Hot or L1, White—Neutral or L2, Orange—Burner motor and Transformer, for constant Ignition. For units that are Intermittent Ignition, Black—Hot or L1, White—Neutral or L2, Orange—Burner motor, Blue—Ignition Transformer. All local electrical codes must be followed.

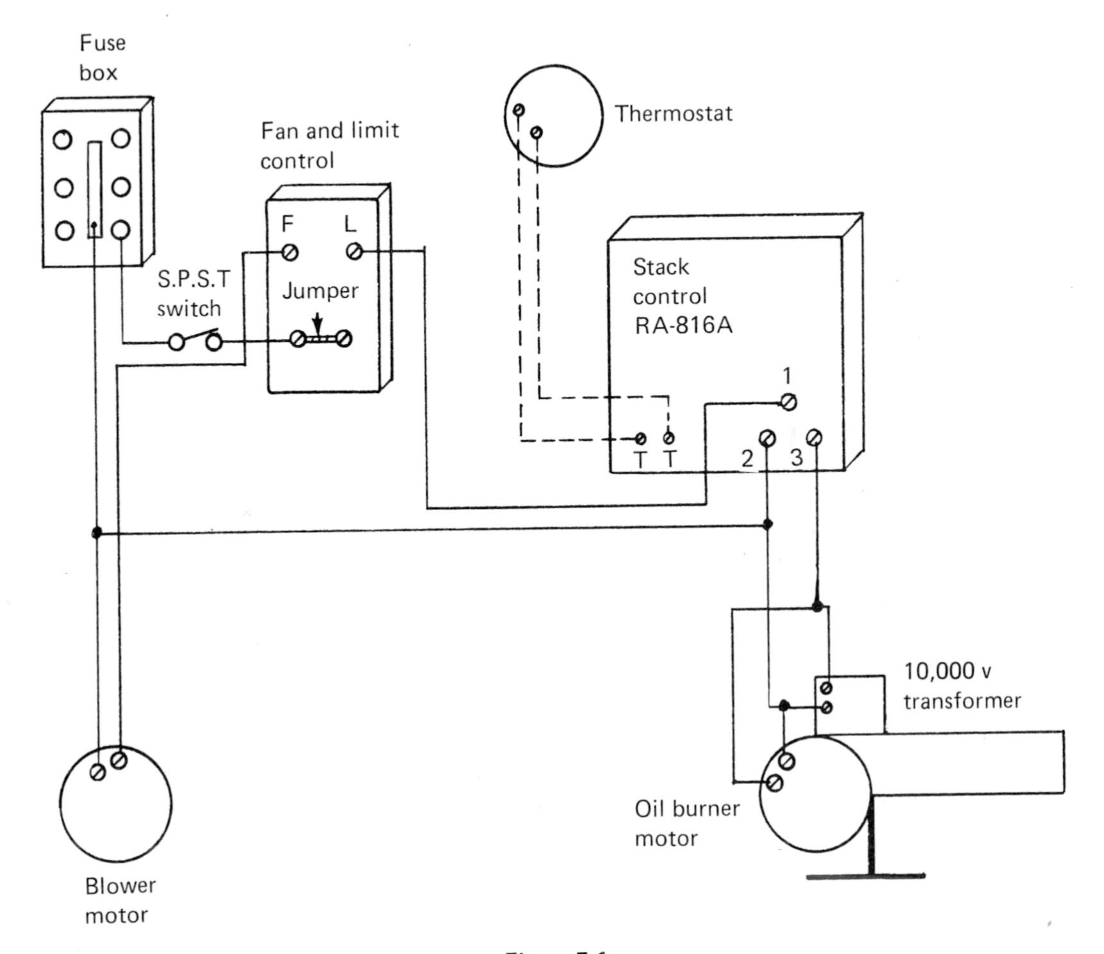

Figure 5.1a.

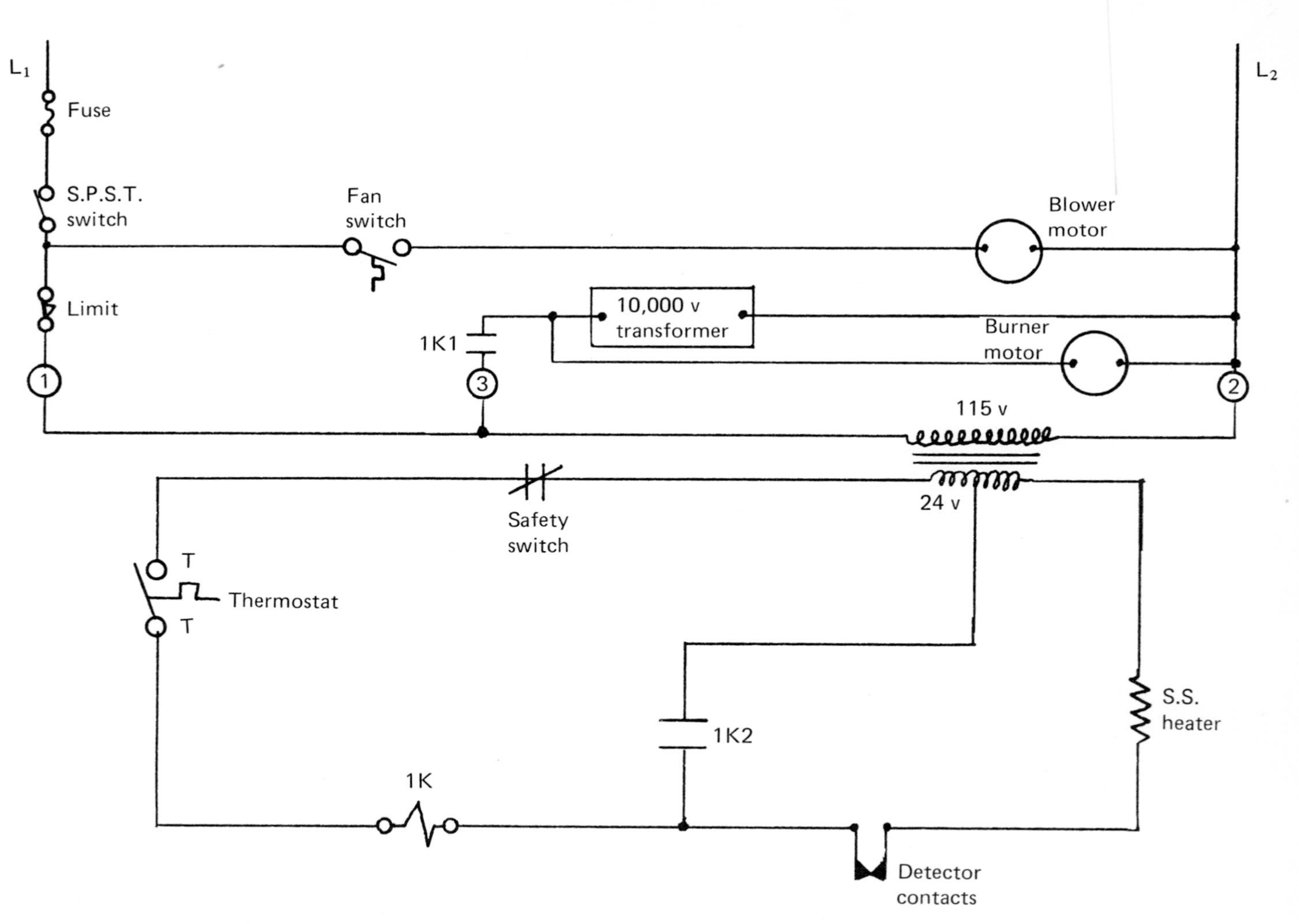

L1
Fuse
S.P.S.T.
switch
Fan
switch
Blower
motor
L2
Limit
10,000 v
transformer
1K1
Burner
motor
1
3
2
115 v
24 v
Safety
switch
T
Thermostat
T
1K2
S.S.
heater
1K
Detector
contacts

Figure 5.1b.

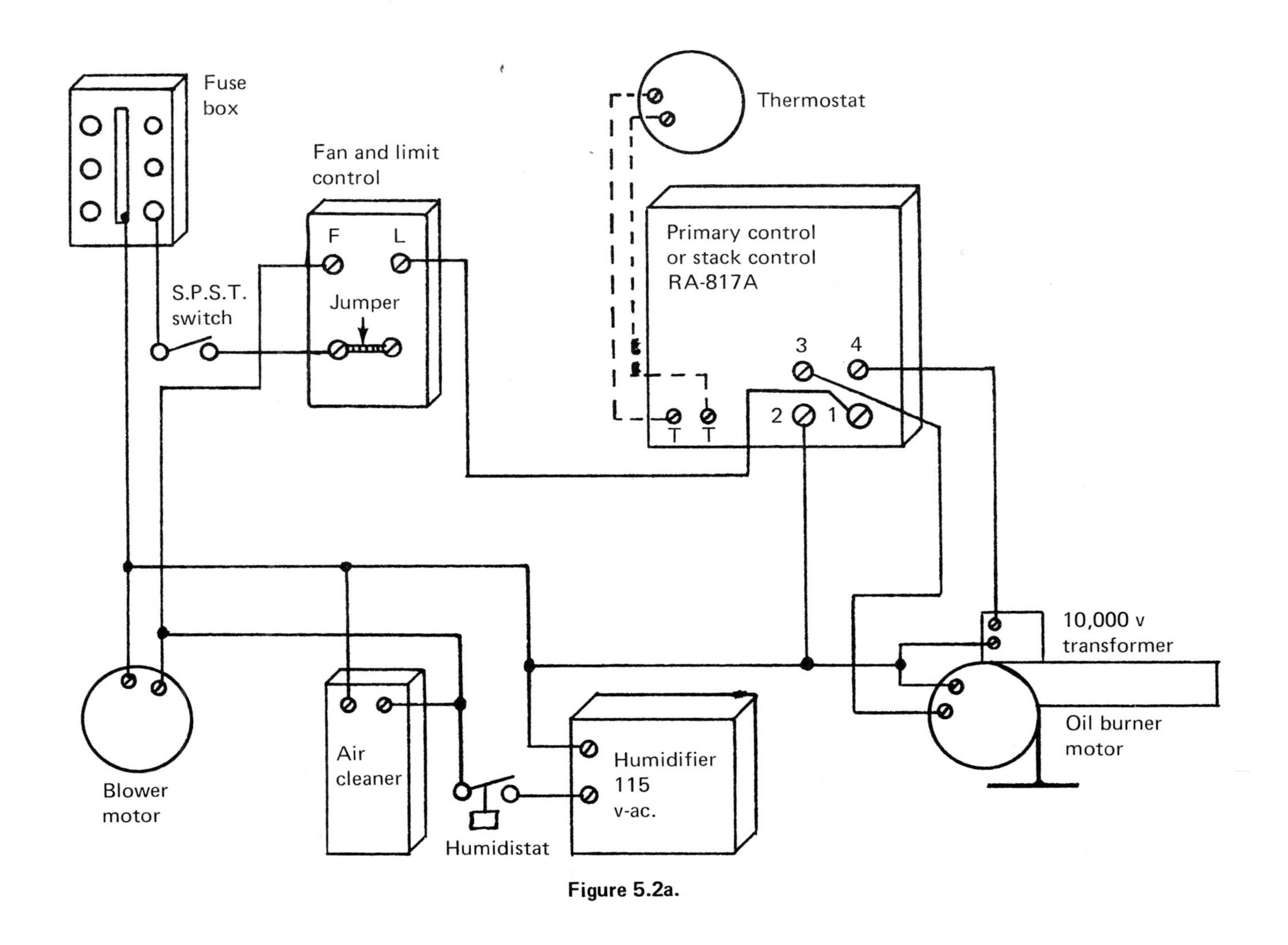

Fuse
box
Fan and limit
control
F
L
Jumper
S.P.S.T.
switch
Thermostat
Primary control
or stack control
RA-817A
3
4
2
1
T
T
10,000 v
transformer
Oil burner
motor
Blower
motor
Air
cleaner
Humidistat
Humidifier
115
v-ac.

Figure 5.2a.

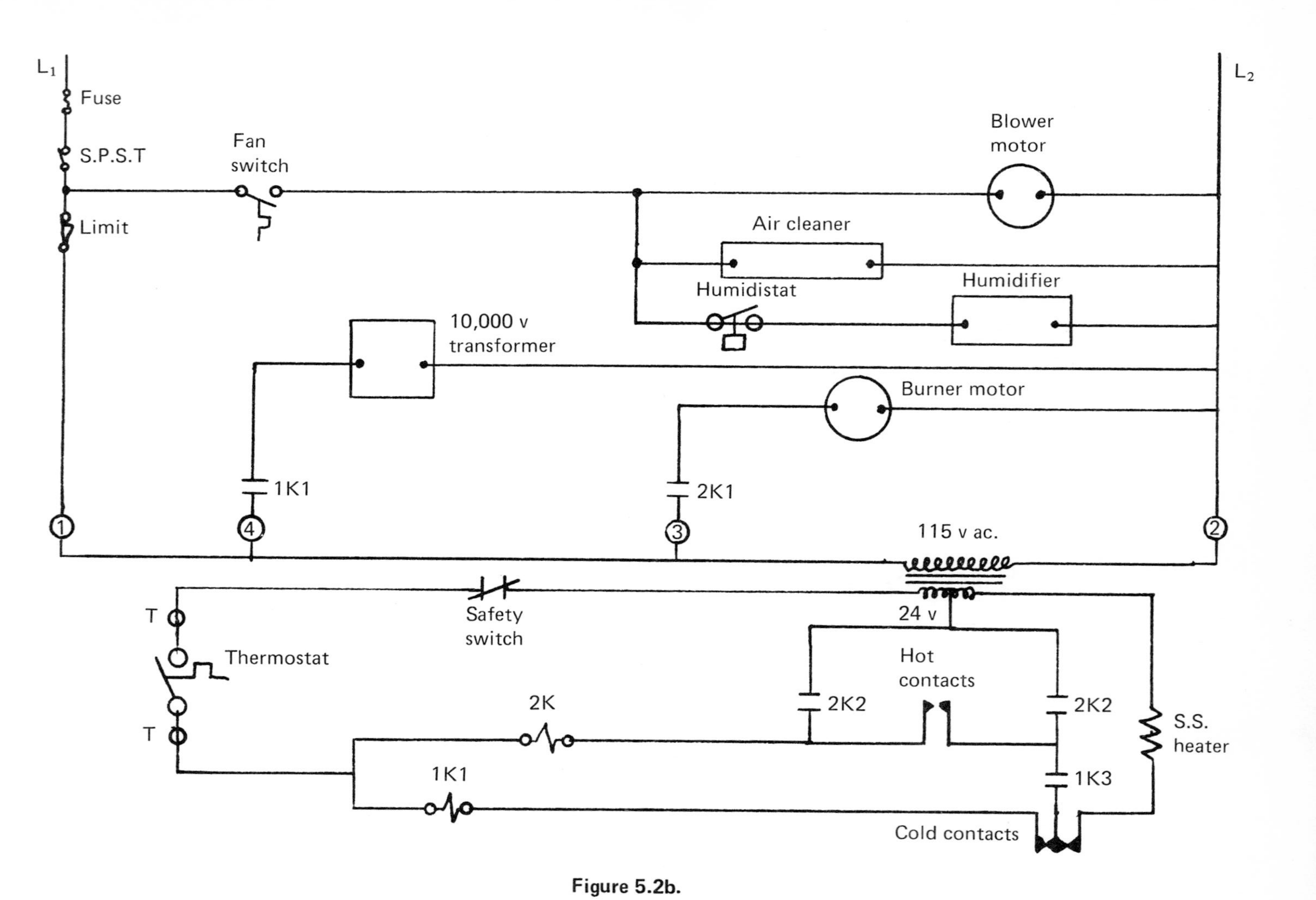

L₁
Fuse
S.P.S.T
Limit
Fan switch
Blower motor
Air cleaner
Humidistat
Humidifier
10,000 v transformer
Burner motor
1K1
2K1
1
4
3
2
115 v ac.
24 v
Safety switch
T
Thermostat
T
Hot contacts
2K
2K2
2K2
S.S. heater
1K1
1K3
Cold contacts
L₂

Figure 5.2b.

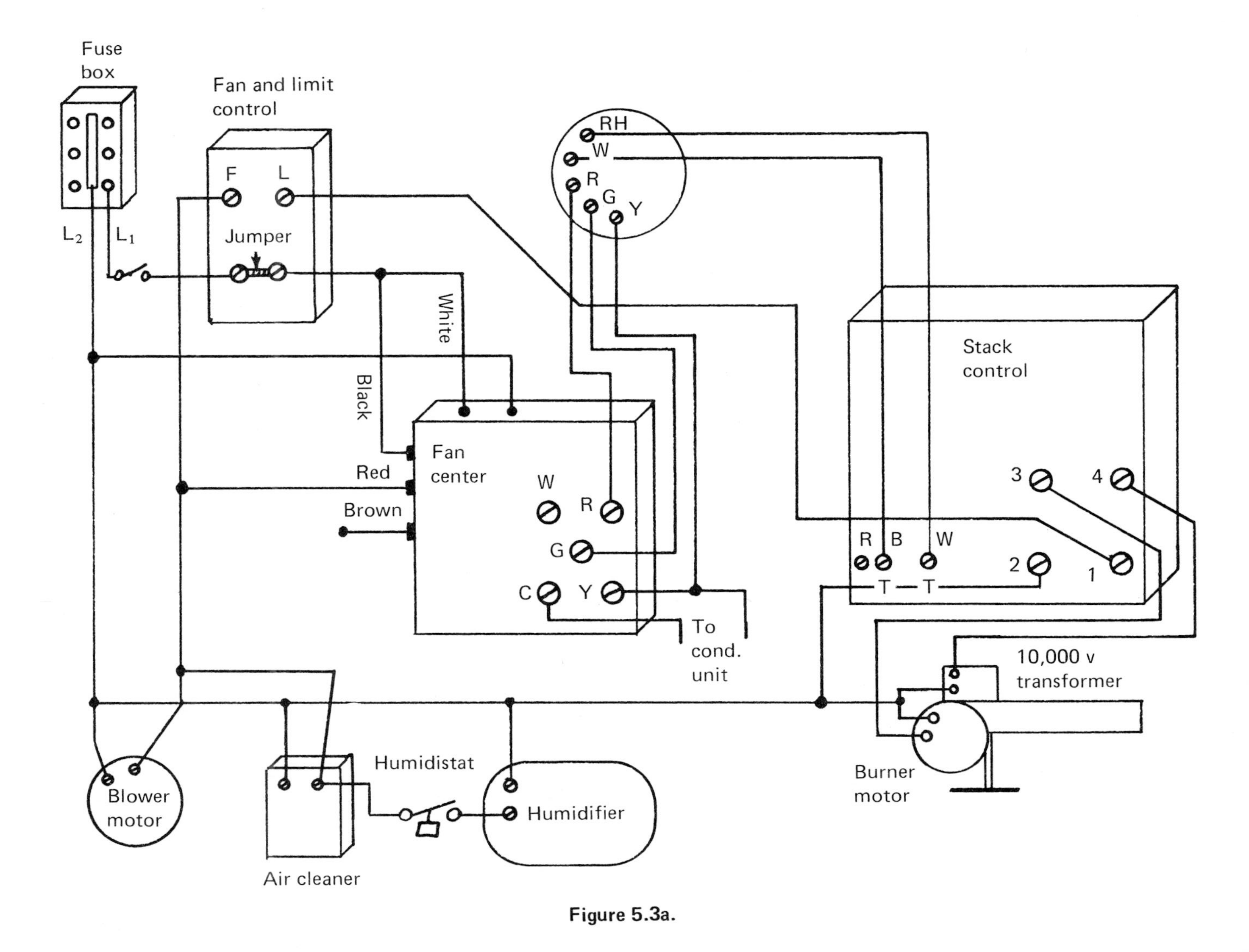

Fuse box
L_2
L_1
Fan and limit control
F
L
Jumper
RH
W
R
G
Y
White
Black
Red
Brown
Fan center
W
R
G
C
Y
To cond. unit
Stack control
3
4
R B W
2
1
T T
10,000 v transformer
Burner motor
Blower motor
Air cleaner
Humidistat
Humidifier

Figure 5.3a.

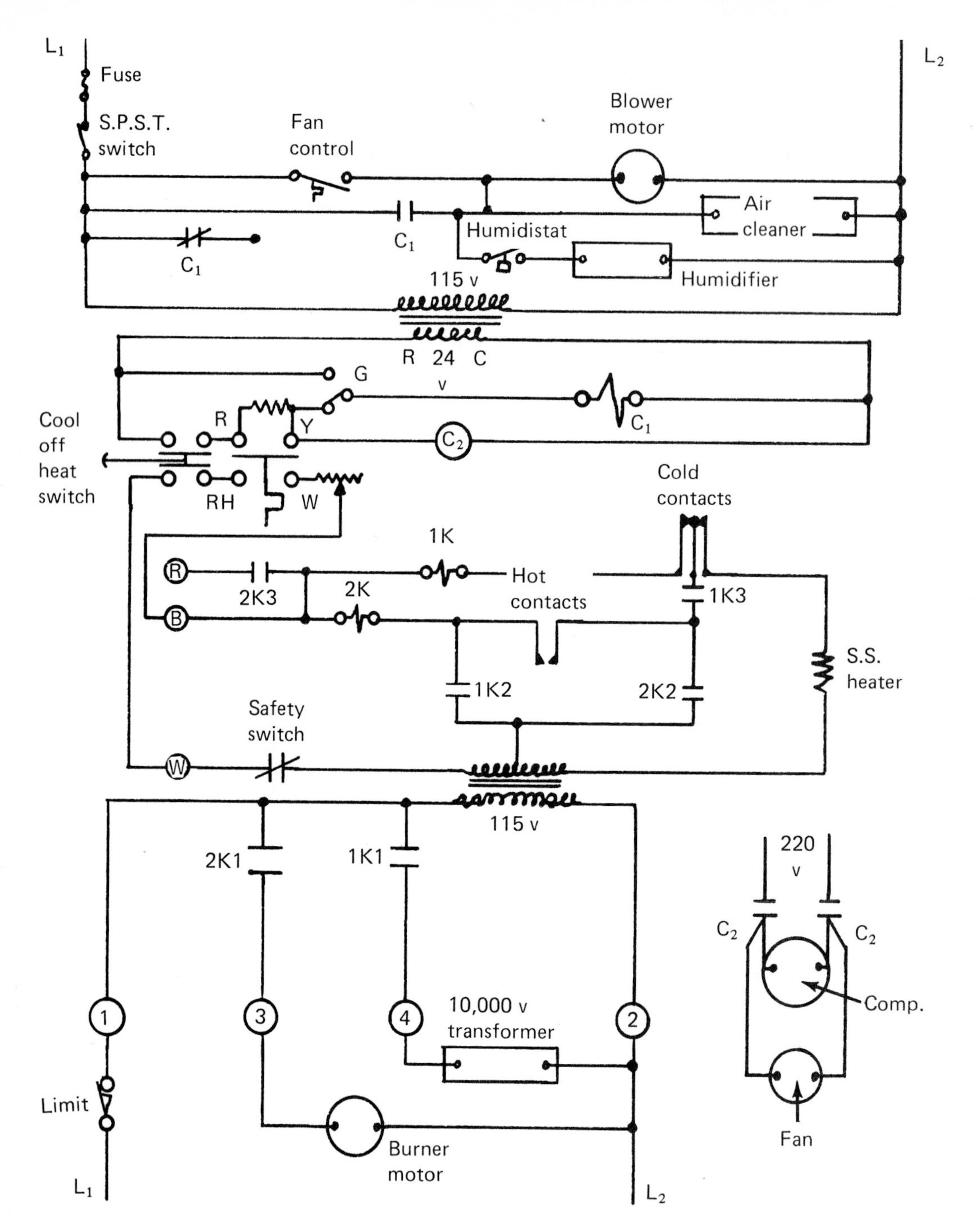
L1
L2
Fuse
S.P.S.T.
switch
Fan
control
Blower
motor
Air
cleaner
C1
C1
Humidistat
Humidifier
115 v
R 24 C
v
G
C1
Cool
off
heat
switch
R
Y
C2
RH
W
Cold
contacts
1K
Hot
contacts
2K3
2K
1K3
S.S.
heater
1K2
2K2
Safety
switch
115 v
2K1
1K1
220
v
C2
C2
Comp.
10,000 v
transformer
Limit
Burner
motor
Fan
L1
L2

Figure 5.3b.

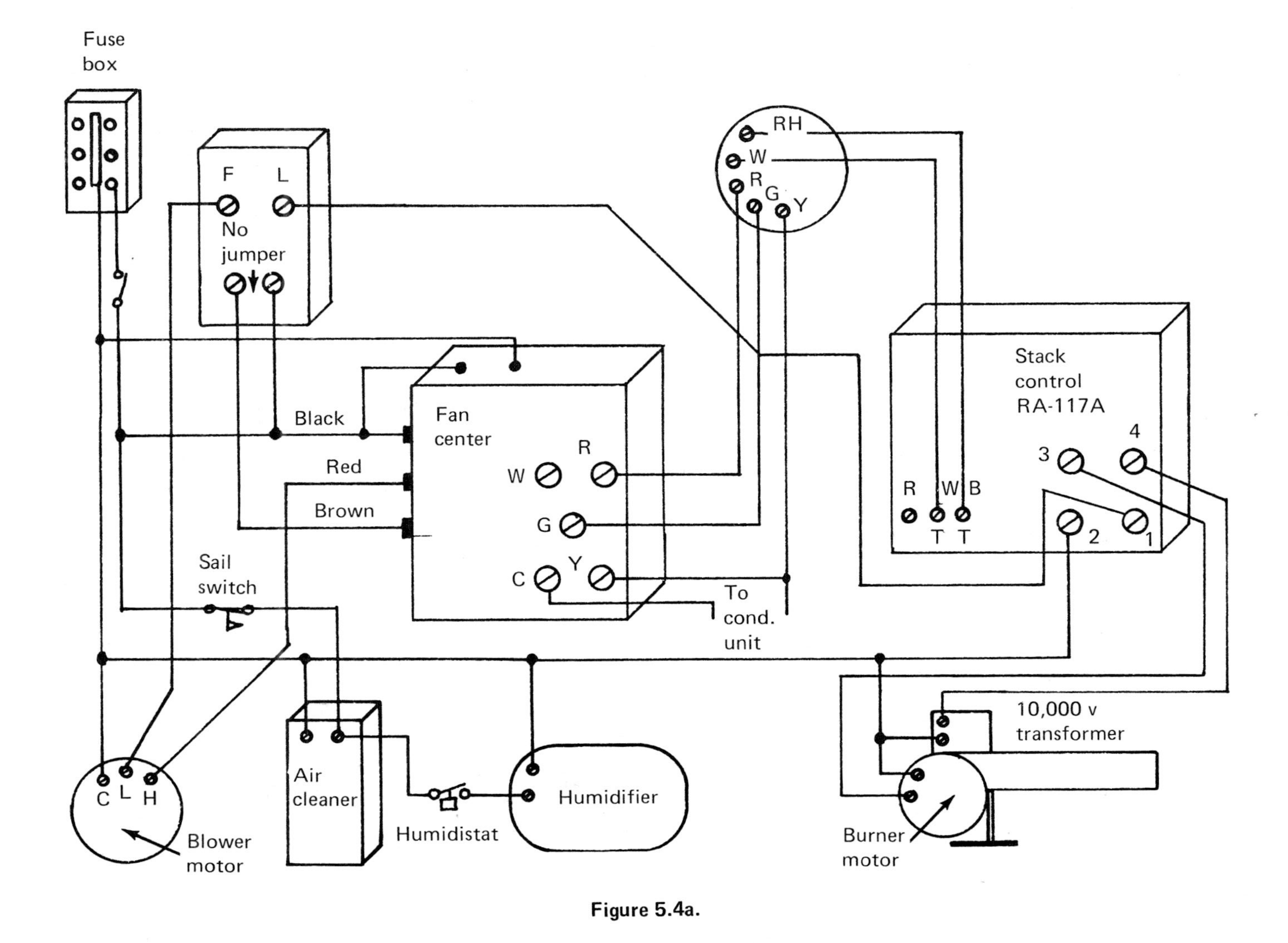

Fuse
box
F
L
No
jumper
RH
W
R
G
Y
Black
Red
Brown
Fan
center
W
R
G
C
Y
Stack
control
RA-117A
4
3
R
W
B
T T
2
1
Sail
switch
To
cond.
unit
10,000 v
transformer
C
L
H
Blower
motor
Air
cleaner
Humidistat
Humidifier
Burner
motor

Figure 5.4a.

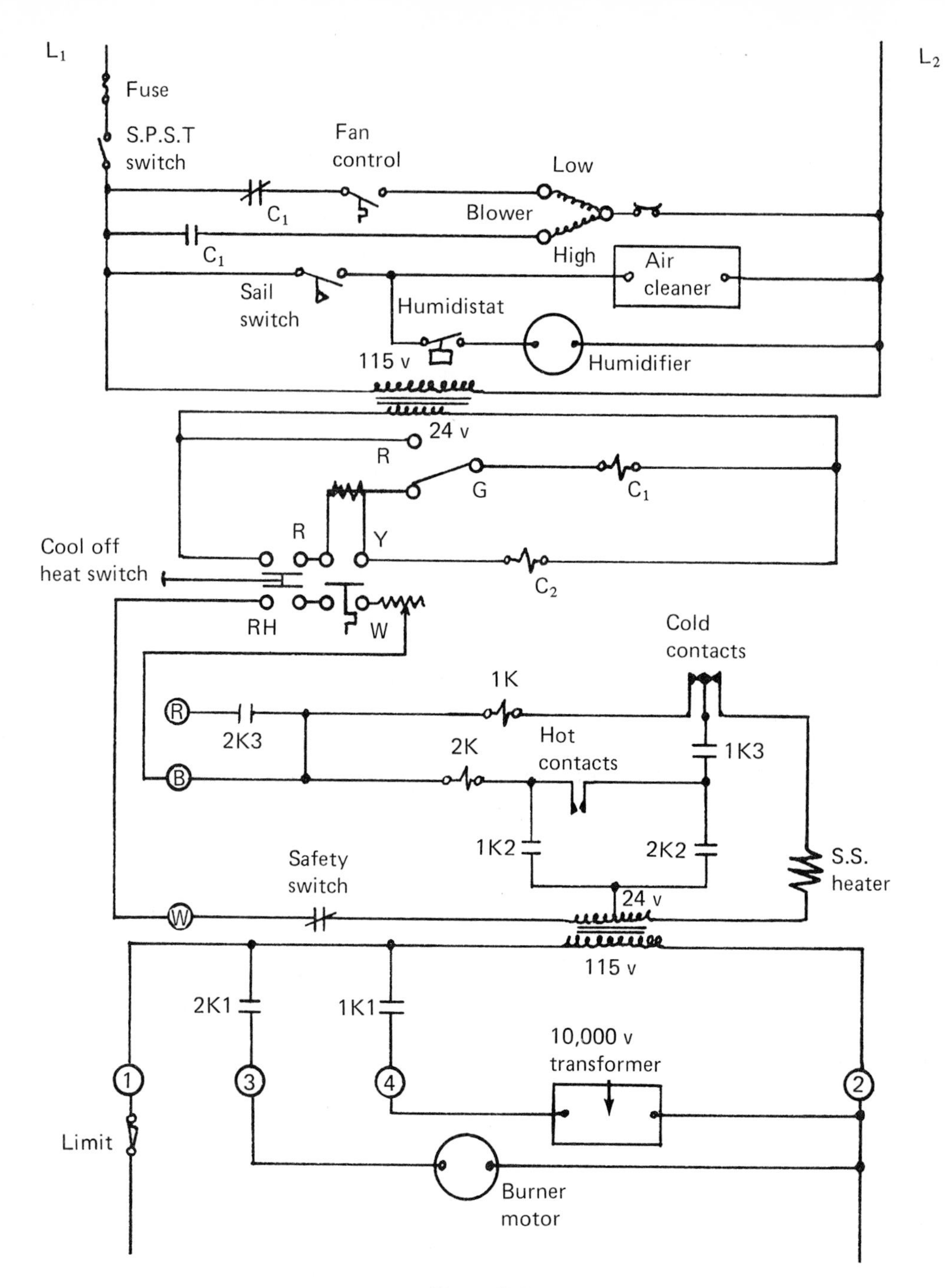

L1
L2
Fuse
S.P.S.T
switch
Fan
control
Low
C1
Blower
C1
High
Air
cleaner
Sail
switch
Humidistat
115 v
Humidifier
24 v
R
G
C1
R
Y
Cool off
heat switch
C2
RH
W
Cold
contacts
1K
R
2K3
1K3
2K
Hot
contacts
B
1K2
2K2
S.S.
heater
Safety
switch
24 v
W
115 v
2K1
1K1
10,000 v
transformer
1
3
4
2
Limit
Burner
motor

Figure 5.4b.

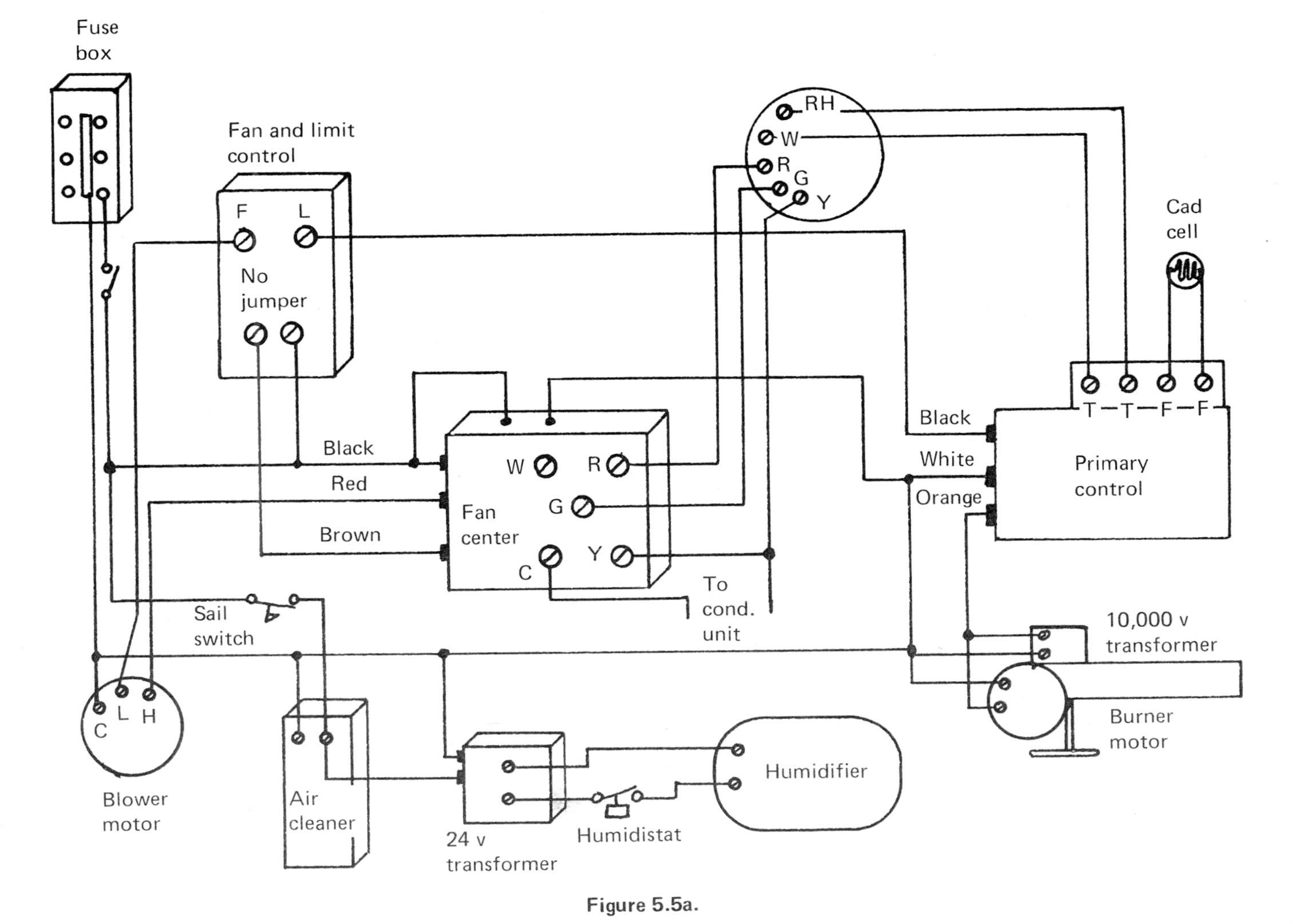
Fuse
box
Fan and limit
control
F
L
No
jumper
Black
Red
Brown
W
R
G
C
Y
Fan
center
To
cond.
unit
RH
W
R
G
Y
Cad
cell
T—T—F—F
Black
White
Orange
Primary
control
10,000 v
transformer
Burner
motor
Sail
switch
C
L
H
Blower
motor
Air
cleaner
24 v
transformer
Humidistat
Humidifier

Figure 5.5a.

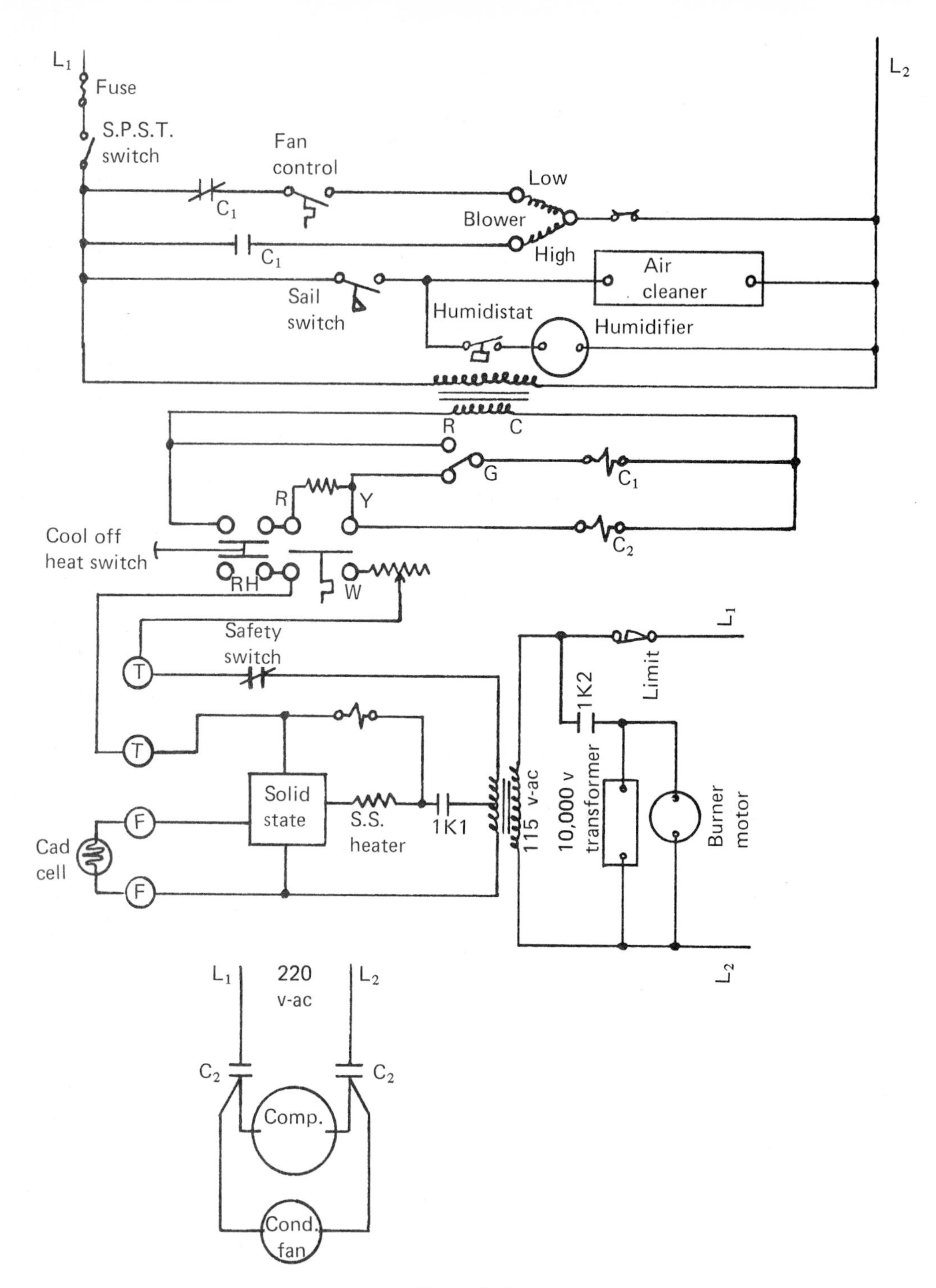
L1
Fuse
S.P.S.T.
switch
Fan
control
Low
C1
Blower
C1
High
Air
cleaner
Sail
switch
Humidistat
Humidifier
R
C
G
C1
R
Y
Cool off
heat switch
C2
RH
W
Safety
switch
L1
Limit
1K2
T
T
Solid
state
S.S.
heater
1K1
115 v-ac
10,000 v
transformer
Burner
motor
Cad
cell
F
F
L2
L1
220
v-ac
L2
C2
C2
Comp.
Cond.
fan
L2

Figure 5.5b.

Chapter 6
FLAME SENSOR CONTROLS

As heating controls have become more sophisticated, various types of flame sensors have appeared, replacing the older bi-metal and diaphragm devices. These all have their own characteristics and are in no way interchangeable.

The cad cell has become increasingly popular for use on residential and light commercial oil burners. It senses light rather than heat. It consists of a shielded grid coated with cadmium-sulphide, whose electrical resistance is extremely high in darkness and very low in the presence of light. In effect, it's a switch, allowing current to pass only when light is sighted. Since the cad cell responds to any light in the visible range, it must be protected from external light sources. It should never be used to detect a gas flame or gas pilot flame. It is not approved at present for large burners. Check your local codes for maximum gallonage allowed. See Figure 6.1.

The ultra-violet detector is another device which acts like a switch, passing or blocking a current according to the presence or absence of certain radiations. However, it will react only to the invisible wavelength of ultra-violet light. As a result, it will respond to an oil flame, a gas flame, and even an ignition transformer spark. It can sight and prove both gas pilot and main flame simultaneously. The advantage of the ultra-violet detector is that it will not respond to any kind of visible light or infra-red. Hot refractory will not activate it, nor will light from outside sources. At present it is used almost exclusively in commercial and industrial systems. See Figure 6.2.

The flame rods are used to prove the existence of a gas-pilot flame for ignition purposes only—on commercial and industrial burners. (Use of a photocell, lead sulphide cell, or ultra-violet detector is necessary for proving the main oil flame.) In this system an insulated probe is mounted so that its tip is in the path of the pilot flame. The gas flame acts both as a conductor and a rectifier, passing on the direct current necessary to operate the control system. See Figure 6.3.

The lead sulphide cell, like the cad cell, is a light-sensing switch whose resistance varies with the amount of light. However, it is more discriminating since it is most sensitive, to certain radiation wave lengths, found in a flame. It reacts to both oil and gas flames and can be sighted so as to prove the pilot and main burner simultaneously. Therefore, it is practical for industrial applications. Most control circuits using a lead sulphide flame detector, incorporate an electronic frequency filter which blocks current flow when a pulsating flame is not present. See Figure 6.4.

The photocell reacts only to orange and red light such as is found in an oil flame. In the presence of this band of light, a specially coated plate within the cell emits electrons and provides a rectified (direct current) path to the grid of a vacuum tube. Here this small current (2 to 5 millionths of an ampere) is amplified sufficiently to energize a relay coil. When the photocell

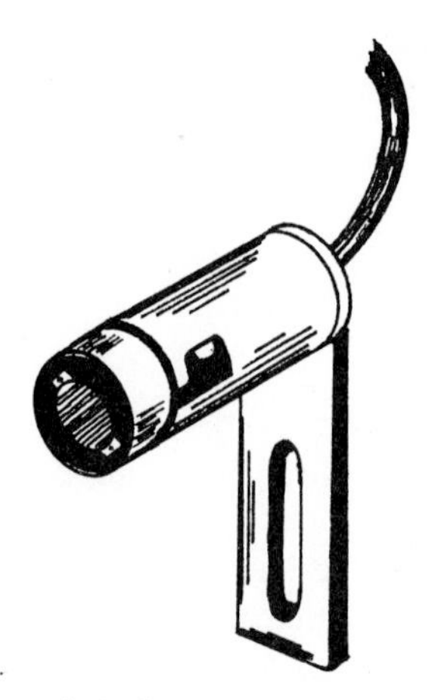

Figure 6.1. Cadmium-sulphide cells.

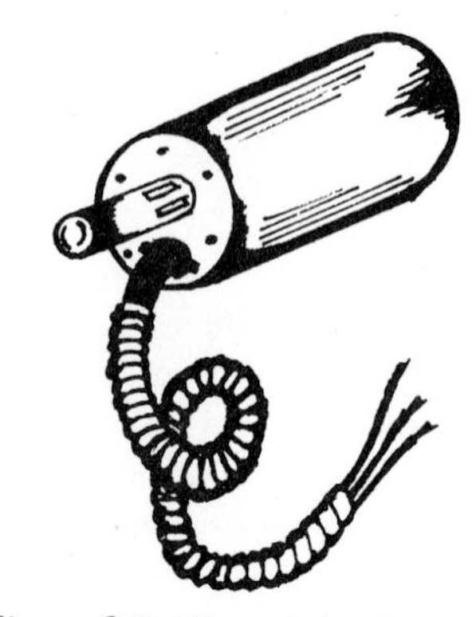

Figure 6.2. Ultra-violet detectors.

Figure 6.3. Flame rod.

Figure 6.4. Lead sulphide cell.

Figure 6.5. Photocell.

cannot view orange or red light, no current flows to the vacuum tube grid. Under this condition the grid completely blocks current flow through the tube and therefore to the coil. The photocell cannot be used to detect a gas or gas-pilot flame. See Figure 6.5.

Cad Cell and Relay

The function of the cad cell relay is to

1. Start the burner on a call for heat.
2. Stop the burner when the heat call is satisfied.
3. Stop the burner when there is a flame failure for lack of oil or proper flame ignition.
4. Stop the burner when power is interrupted due to a limit trip.

The three most commonly seen cad cells and relays, are Honeywell, White-Rodgers and Simicon. They are all somewhat different in appearance, and are also different in their internal components. All operate with the same type of sensing element, the cad cell. The cad cell performs the same basic function in each case, which is to sense either light or darkness. If you were to look at the face of a cad cell, you would see a line zigzagging back and forth across this face, this line is cadmium sulfide. Cadmium sulfide is a material that has a high resistance to the flow of electricity in the dark, but has a low resistance in bright light. When checking the cad cell you should use an ohm meter for checking. You first should determine which scale or position you desire. You want to select the scale that will show an easy to read meter movement in the 1,600 ohm or less range. After you have selected the correct position of the scale and put one end of each lead into the correct places in the meter case then put the loose ends of the leads together, this will cause the meter hand to go to zero. Turn the zero adjustment knob until the hand is exactly on zero. Take the leads apart and again touch them together. The meter hand should go full scale towards the high reading and back to zero when the leads are touched together. Now connect the leads to the wires of a cad cell you have removed from a burner. You will notice the meter will show a very low reading when the cell is exposed to light. Cover the end of the cell with your thumb, and the reading will go to a very high resistance. This action is the same as is experienced in a burner.

In the off position, the cell is in the dark and has a high resistance. When a proper flame is seen, the resistance drops down into the low range of 1,600 ohm or less, normally in the 400-1,200 ohm range. A properly operating cad cell and relay will operate in this manner. On a call for heat, the cell has a very high resistance because it is dark, 50,000 ohms or more. As the burner ignites and burns, this resistance will normally drop to 400-1,200 ohms range. Should the burner fail to ignite, the relay will go off on safety. The reason it does is because of the safety heater of course. The cad cell is what makes this happen in this manner. The internal circuit of the relay is such that the power to the safety circuit has two places it can go, either to the safety heater or to the cad cell. If the cad cell sees enough light due to its low resistance, current would rather flow through it than through the higher resistance of the safety heater. The safety heater gets no current flow, so the burner does not go off on safety. Should there be a flame failure, the resistance of the cell goes up immediately, and with its now high resistance, the current will go through the safety heater rather than the cell, so in a short period of time, 30-60 seconds, the safety circuit shuts down the burner. See Figure 6.6.

Figure 6.6. Honeywell relay.

To check the operation of the relay the following steps should be taken:

1. With the burner off, remove both cad cell leads from the relay.
2. Connect one end of a short piece of wire to one of the relay terminals, where the cad cell wire was connected. Leave the other end of the wire loose.
3. Zero the ohm meter as described earlier. Then connect the meter leads to cad cell leads. One lead to each wire of the cad cell. The meter should show a very high resistance. 50,000 ohms or more.
4. Start the burner. As soon as the flame is established, quickly fasten the loose end of the jumper wire to the other relay terminal (where the second cad cell lead had been). Do this as quickly as possible, or the relay may go off on safety.
5. The burner should continue to run. If it does run properly for five minutes, the relay should be considered to be all right.
6. With the burner running, and a flame, now read the ohm meter. This should be in the 1,600 ohm or less range. If the resistance is higher, then:
   a. The cell is dirty.
   b. The cell may be somewhat out of position.
   c. The cell could have a scorched face from operation on a poor draft situation.
   d. It is also possible that the cell contacts in the socket are corroded.
   e. You may have a bad cell. A defective cell can be determined by checking its resistance when exposed to a bright light, and with the face covered.

Resistance should be 100 ohms or less when exposed to a bright light, 50,000 or more when covered over so it can see no light. Light will shine through your thumb, so do not use your thumb to cover the cell to stimulate darkness.

7. Should the burner stop for any reason during the test, remember you must remove one end of the jumper wire from the relay, or the relay will not pull in again. The reason of course, is that these terminals are not normally jumpered. They are connected to the cell with no flame (burner off) the cell is in the dark so it has a high resistance to current flow. By removing one end of the jumper wire from the relay, you are stimulating this high resistance of the dark cell.
8. Again, if all was well as explained in No. 5, but the burner had been going off on safety in the past, the ohm resistance check will in all probability show high resistance. This is the probable cause of these safety shutdowns. You can only confirm this by a resistance check as was explained. There must be 1,600 ohms or less resistance during burning operation, with 400-1,200 recommended. It must be less 1,600 or nuisance lockouts will occur at times.
9. Do not overlook the possibilities of low voltage. Voltage of 107v or less will cause control problems, as well as the possibility of overheating of the burner motor and poor ignition.
10. It is always a good practice to simulate a flame failure by shutting off the oil flow, removing one cad cell lead, and opening the limit circuit to check all possible safety functions. These are three separate tests that will verify that all safety functions and controls are in order.

Pressure Regulators—Natural Gas

Natural gas pressure at the street will be from 15 lb. to 60 lb. After the regulator on the gas meter the pressure is generally distributed between 6″ W.C. to 7″ W.C. line pressure. This pressure will vary depending on the load. The firing rate of a furnace depends on the manifold pressure which on Natural Gas is 3 1/2″ W.C. Therefore, a pressure regulator is required to reduce the 6″ W.C. to 3 1/2″ W.C. and maintain this pressure on the manifold as the line pressure varies.

The pressure regulator is actually an automatic gas valve. During the non-operation part of the cycle when no gas is being used pressure in the piping between the regulator and the gas valve build up to full static pressure. This build-up raises the diaphragm until the regulator valve is fully closed. When the operating cycle begins, the gas valve opens and the pressure on the outlet side drops, the diaphragm spring pushes the diaphragm down and opens the valve mechanism. During the on-cycle a balance is maintained between the lift power of the gas under the diaphragm and the spring tension above. The pressure is regulated by adjusting the spring tension.

1. Never attempt to set pressures by visual flame observations. Always use a "U" gauge or a mechanical gauge calibrated in inches. Set to manufacturers specifications.
2. Operating pressure is the only true pressure and should be taken with burner in operation. Never accept static pressure as a satisfactory operating pressure.
3. Always check flow direction when installing a pressure regulator. Proper position is usually indicated by an arrow on the bottom of the casting. A regulator cannot function if installed in a reversed position.

4. Correct pressure is important to proper combustion. If pressure is lower than required, the air volume inspirated into the venturi will be reduced and may result in a yellow tipped flame. Excessive pressure may result in combustion noise and/or flame lifting off the burners.
5. Always check capacity of a regulator before installation. Use manufacturers "spec. sheets."

Service of Honeywell Solid State Controls

Consisting of Thermostat T-7048A, Sub base Q-539M-1011, Interface K-7238B-1004 and two stage gas valve.

This series of checks should be performed on all service pertaining to the following:

1. No Heat Call
2. No Low Fire
3. No High Fire
4. No Blower
5. No Cooling

Reference will be made to various terminals for checking purposes. Pay particular attention as to whether the terminal has a ‾ across the top of the letter. These terminals so indicated are D.C. voltage; the plain lettered terminals such as R, C, or Y are 24 volt A.C. terminals.

Figure 6.7. Honeywell solid state thermostat.

Figure 6.8. Honeywell solid state subbase #Q539M-1011.

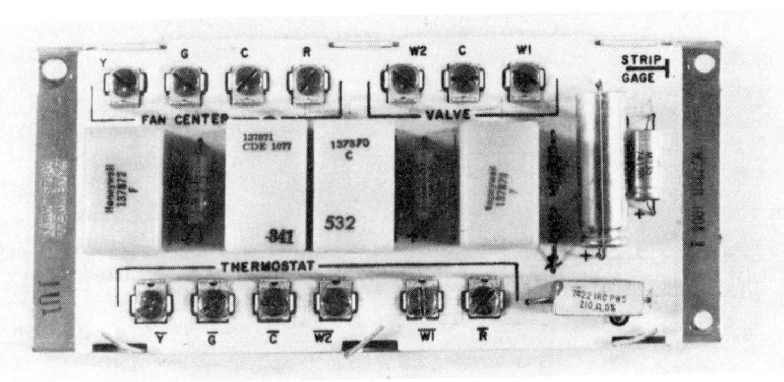

Figure 6.9. Honeywell interface.

During the following description of this System, we will be referring to a device called an "*Interface*" control. This device in addition to being a terminal board for the A.C. and D.C. wiring, also provides the A.C. to D.C. rectification and the pilot duty D.C. Relays which switch the A.C. circuits for Burner, Blower and Cooling Contactor.

1. Place the D.C. Volt Meter across Terminals $\overline{\text{R}}$ and $\overline{\text{C}}$ of the Interface. Reading should be 16—17 Volt D.C. If this voltage is available;
   1. You have 120 Volt Power Supply
   2. The Fan Center Transformer is okay
   3. The D.C. Power Service is okay

2. If there is no D.C. reading at $\overline{R}$ and $\overline{C}$ then proceed to check;
   1. R-C terminal on top of Interface for 24 volts A.C.
   2. If there is 24V at R-C then the Interface is defective and must be replaced. (Note: Identify wires on top and bottom of Interface before replacing.)

If there is no 24V reading at R-C on Fan Center—check for 120V at Transformer Primary. If 120V is available the Transformer in the Fan Center is defective and entire Fan Center must be replaced.

If no voltage is available, check the Limit Control for an open circuit, or the Main Power Supply to unit.

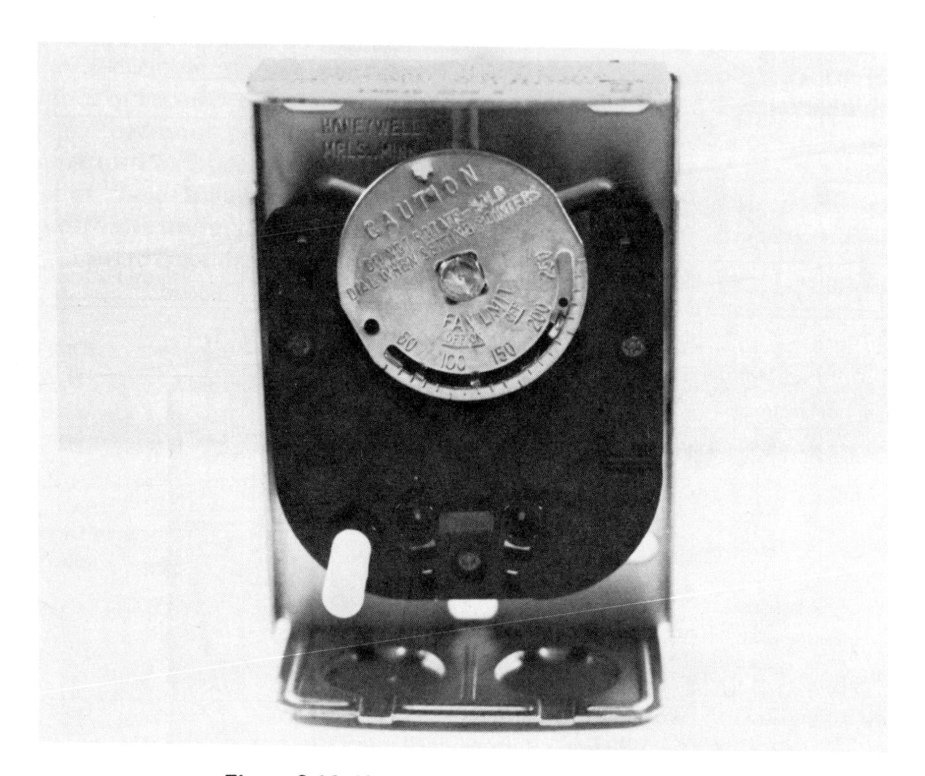

Figure 6.10. Honeywell fan and limit control.

No Heat Call

1. **The previous electrical check out should be made before proceeding.**
2. Jumper $\overline{R}$—$\overline{W}_1$ on bottom of Interface control, and burner should go on low fire. If it doesn't check wiring between Interface and Valve terminals. If wiring is okay valve is defective and must be replaced.
3. If the valve does come on low fire—leave $\overline{R}$—$\overline{W}_1$ jumper in place and add a jumper between $\overline{W}_1$ and $\overline{W}_2$ ***(Caution: Do not allow jumper to touch C terminal).*** In 30 to 45 seconds burner should come on high fire. If the burner does not respond to high fire—replace entire valve or operator section only.

When system is calling for first Stage Heating (Low Fire), valve terminals W_1 and C should show 24 volts A.C. with an A.C. Meter.

When system is calling for second stage heating (High Fire), valve terminals W_2 and C should show 24 V, A.C. with an A.C. Meter, and W_1 and C will also show 24 V.A.C.

If burner responds to high and low fire at the interface but will not function from the thermostat, proceed to:

4. Remove thermostat from sub-base. Set sub-base selector to Heat position. Jumper R and W_1 on sub-base, and burner should come on low fire. Perform step #5 below for high fire check. If the burner comes on there is a problem between the thermostat and the sub-base. Replace the thermostat, or sub-base. If burner does not come on check wiring between Interface and thermostat sub-base.
5. With R and W_1 jumpered, add a jumper from W_1 to W_2 and the burner should come on high fire within 20 to 30 seconds. If it does, replace the thermostat or sub-base as in step #4.

No Blower Operation

1. Check $\overline{R}$—$\overline{C}$ on bottom of Interface for 16—17 Volts D.C. If this reading is okay, you have proved electrical power check for A.C. and D.C. Proceed to step 2.
2. Manually depress Contactor in Fan Center. When Honeywell R8239 Relay is used this step must be by-passed. *Motor does not run*—Replace Motor. *Motor does run*—Proceed to step 3.
3. Jumper R—G in Fan Center—Contactor should pull in—*Motor does not run*—Replace Fan Center Contactor—*Motor runs*—proceed to step 4.
4. Jumper R—G on top of Interface. *Motor does not run*—check wiring between Fan Center and Interface. *Motor runs*—proceed to step 5.
5. Similarly jumper $\overline{R}$ and $\overline{G}$ at bottom of Interface. *Motor does not run*—Replace the Interface. *Motor runs,* proceed to step #6.
6. Remove thermostat from sub-base. Jumper R-G. *Motor does not run,* Check wiring between sub-base and Interface. *Motor runs,* Replace stat or sub-base as necessary

1. Check $\overline{R}$—$\overline{C}$ on bottom of Interface for 16—17 V.D.C. If this reading is okay we have proved electric power check for A.C. and D.C. Proceed to step 2.
2. Install a Jumper between R—Y on the Fan Center. *Cooling contactor does not Pull in*—check low voltage wiring for 24 V at contactor. If voltage is present replace contactor. *Cooling Contactor does pull in*—Proceed to step #3.
3. Remove the jumper from step #2 and install it at top of Interface between R and Y. *Contactor does not pull in* check wiring between Interface and Fan Center. *Contactor does pull in,* Proceed to step 4.
4. Remove Jumper from top of Interface and Install between $\overline{Y}$ and $\overline{R}$ at bottom of Interface. *Contactor does not pull in* Replace the Interface. *Contactor does pull in,* Remove Jumper from bottom of Interface and proceed to step 5.
5. Remove thermostat from sub-base. Jumper Y to R. *Contactor does not pull in*—Check the wiring from the sub-base to Interface. *Contactor does pull in,* Replace thermostat or the sub-base as required.

Belt Drive Blowers with Variable Pitch Motor Pulley

Mount the variable pitch motor pulley on the motor shaft with its movable face toward the end of the shaft, away from the motor. On the side of the pulley nearest the motor. Tighten the set screw on the "flat" or "key" of the motor shaft. Leave some clearance between the pulley and the end belt of the motor. Wipe off pulleys and belt with a clean rag to get rid of all oil and dirt. Dirt and grease are tough abrasives that cause the belt to wear out faster. When installing a v-belt in pulley grooves loosening the belt take-up or the adjusting screw on the motor. Do not "roll" or "snap" the belt on the pulleys. Make sure the belt doesn't bottom in the pulley grooves. The alignment of both pulleys and shafts is accomplished by moving the motor on its mount. You can do this "by eye" but you're a lot safer if you hold a straight edge flush against the blower pulley, then move the motor until the belt is absolutely parallel to the straight edge. Run the motor to determine the direction of rotation. If necessary to reverse the rotation, follow the instructions on the motor terminal block for reversing the lead wires. The directions are usually found under the cover where the lead wires enter the motor. Run the blower for about five minutes, or until the belt tends to "seat" itself in the pulley grooves. To adjust the blower speed, increase or decrease its output of air in a furnace, loosen the set screw on the outer face of the variable pitch motor pulley and turn the face. The speed is reduced by turning the outer face so as to move the two pulley faces farther apart. The speed is increased by turning the outer face so as to move the faces closer together. Retighten the set screw against the flat spot on the pulley hub after adjusting the speed of the blower.

Adjusting Blower Speed in a Forced Warm Air Furnace

Seventy degrees Farenheit is a recommended heat rise unless otherwise specified by the manufacturer.

Place one thermometer in the return air plenum, and one in the warm air plenum. After the temperature of the warm air side has leveled off with the blower and burner running compare the readings of the two thermometers. If the difference is less than 70 °F, slow down the blower speed. If it is more than 70 °F, increase the blower speed. Keep adjusting until a rise of approximately 70 °F is maintained.

In air conditioning unit, use a 20 °F drop instead of the 70 °F rise for heating. (Unless otherwise specified by the manufacturer of the unit.)

Air Flow Formulas

Gas Furnace

$$\text{CFM} = \frac{\overset{\text{Heat Value of Gas}}{(\text{BTU/CU.FT.})} \times \text{CU. FT./HR.} \times 0.8}{1.08 \times \text{TEMP. RISE*}}$$

or

$$\text{CFM} = \frac{\text{BTU INPUT} \times .8}{1.08 \times \text{TEMP. RISE*}}$$

Oil Furnace

$$\text{CFM} = \frac{\overset{\text{Heat Value of Oil}}{(\text{BTU/GAL})} \times \text{GAL/HR.} \times .8}{1.08 \times \text{TEMP. RISE*}}$$

or

$$\text{CFM} = \frac{\text{BTU INPUT} \times .8}{1.08 \times \text{TEMP. RISE*}}$$

Electric Furnace

$$\text{CFM} = \frac{\text{VOLTS} \times \text{AMPS} \times 3.4}{1.08 \times \text{TEMP. RISE*}}$$

or

$$\text{CFM} = \frac{\text{KW} \times 3{,}410}{1.08 \times \text{TEMP. RISE*}}$$

Examp. of Gas

$$\text{CFM} = \frac{\text{BTU INPUT} \times .8}{1.08 \times \text{TEMP. RISE}} =$$

$$\frac{120{,}000\ \text{BTU} \times .8}{1.08 \times 90^\circ\text{F}} = \frac{96{,}000}{97.2} = 988$$

988 CFM

*Difference between supply air and return air temperature.

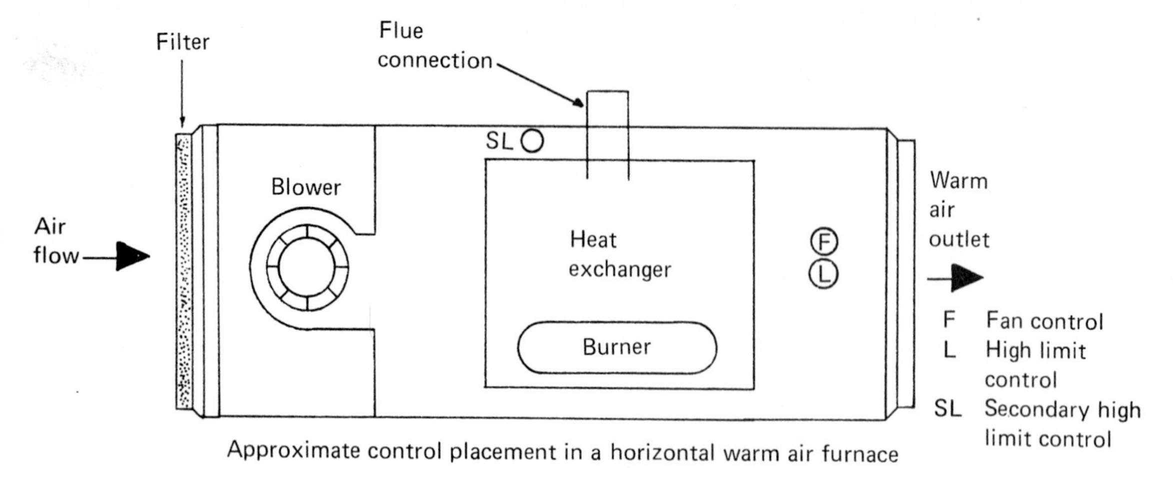

Approximate control placement in a horizontal warm air furnace

Figure 6.11a. Typical locations for fan and limit control.

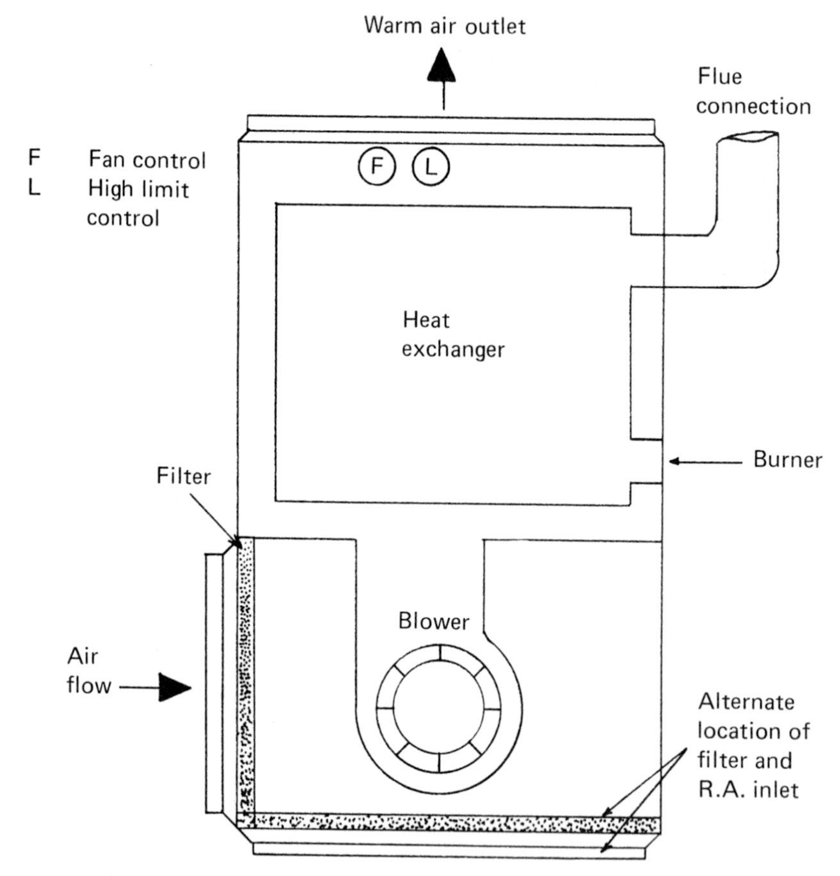

Approximate control placement in an upflow, or high-boy warm air furnace.

Figure 6.11b. Typical location for fan and limit control.

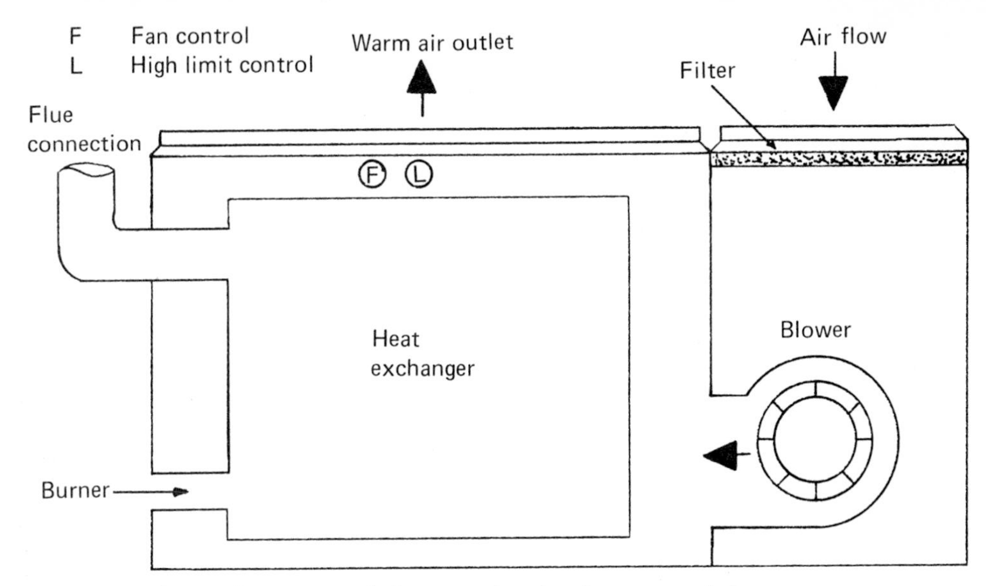

Approximate control placement in a low-boy warm air furnace.

Figure 6.11c. Typical location for fan and limit control.

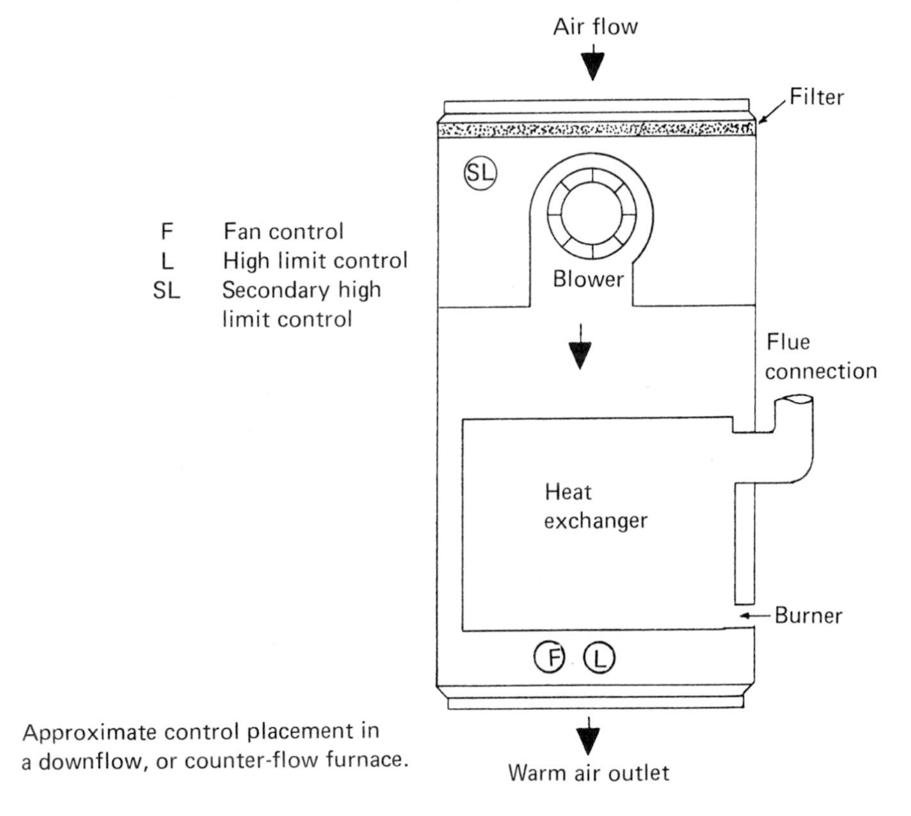

Approximate control placement in a downflow, or counter-flow furnace.

Figure 6.11d. Typical location for fan and limit control.

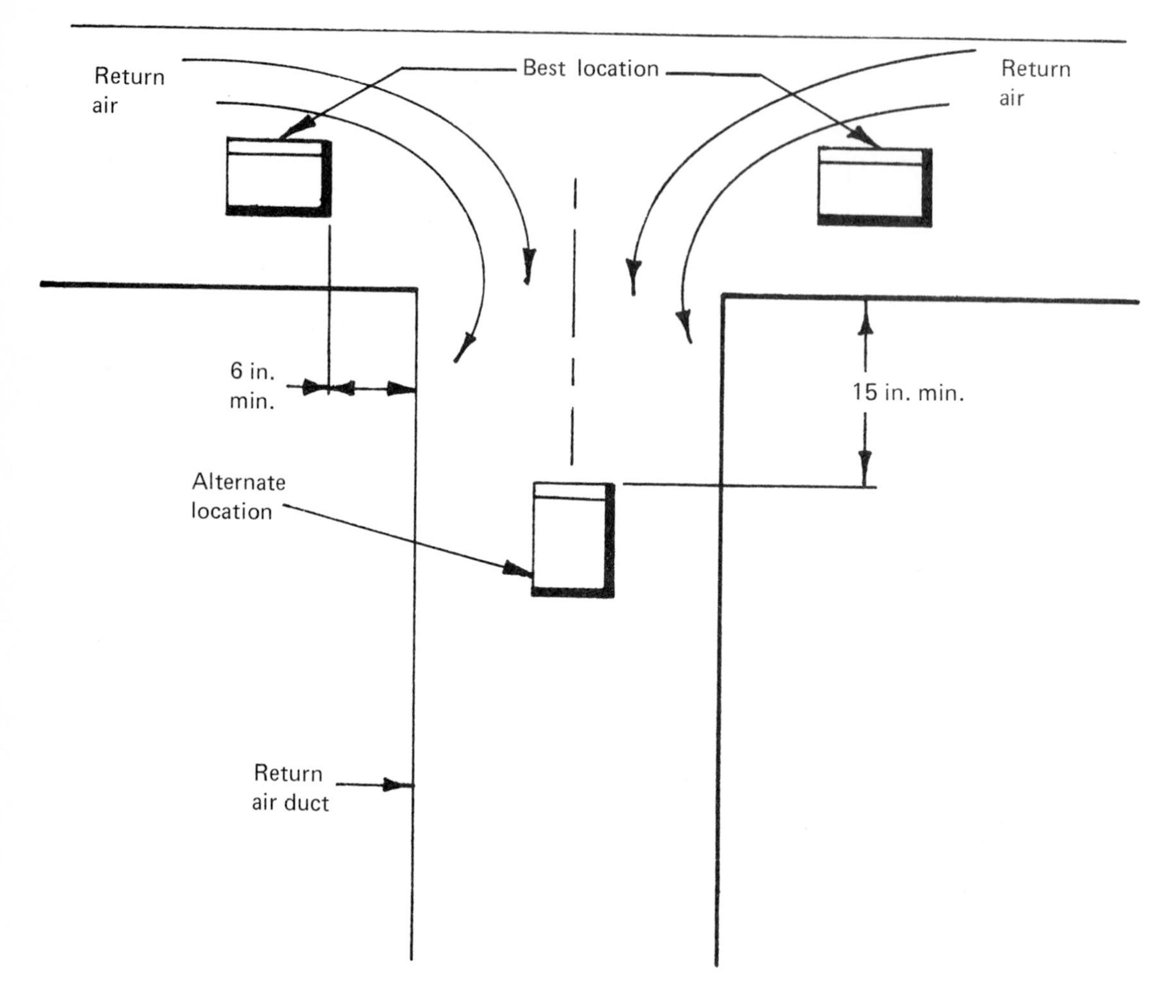

Figure 6.12a. Typical location for air-switch.

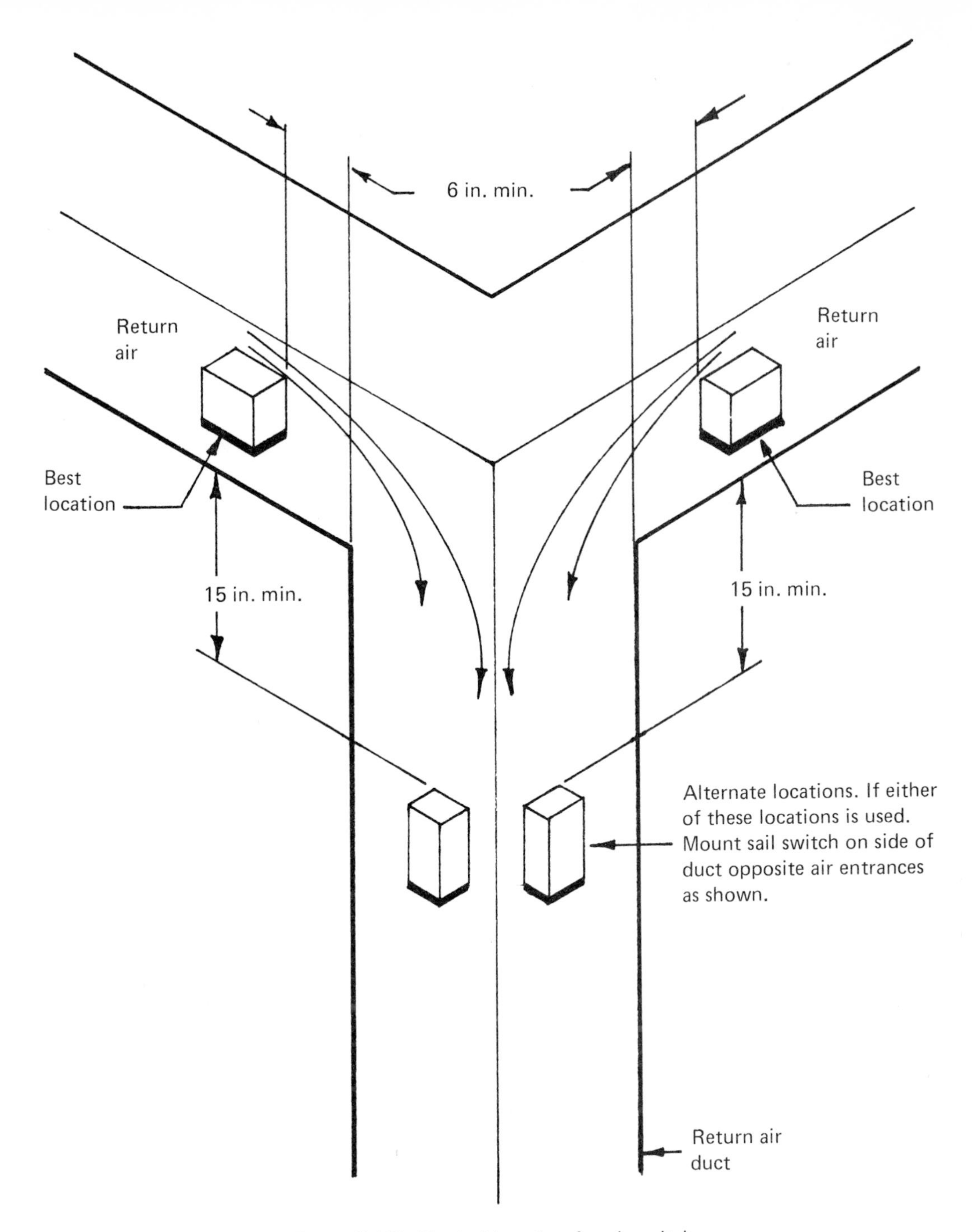

Figure 6.12b. Typical location for air-switch.

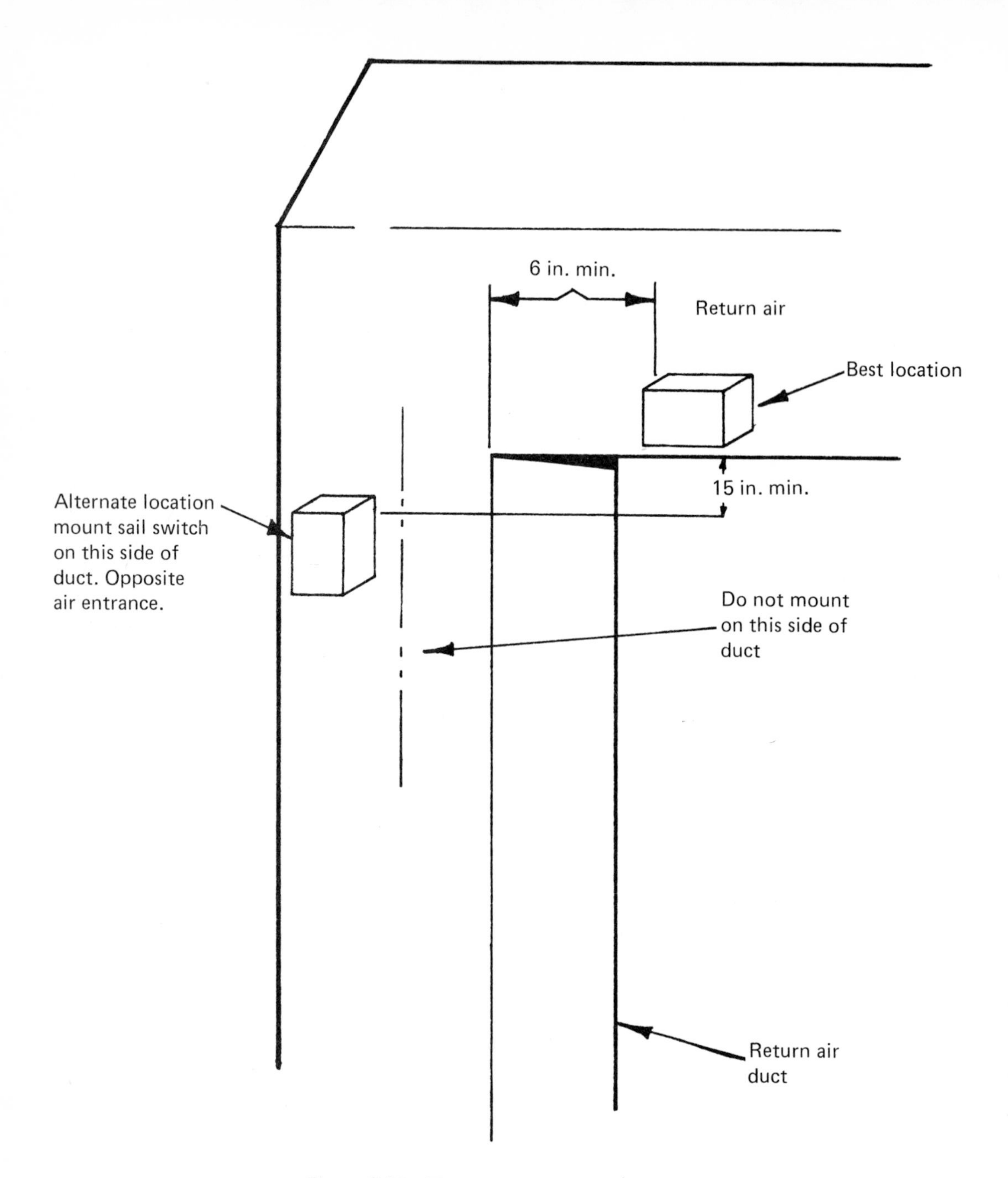

Figure 6.12c. Typical location for air-switch.

Chapter 7

AN INTRODUCTION TO GAS

There are different theories about the origin of natural gas. It is generally thought that natural gas was formed from the decomposition of plant life buried in the earth millions of years ago. Since these plants lived at the same time as those that are now found as fossils, natural gas is called a fossil fuel. Often natural gas and oil are found together in the same well, however the well may have gas or oil by itself. Natural gas, like oil, is a mixture of hydrocarbons. Both are fossil fuels composed of various chemicals derived from the hydrogen and carbon contained in prehistoric plants. The chemical components of natural gas are: methane about 85%, ethane about 12%. The remaining 3% is made up of traces of other chemicals. When we take raw natural gas from the earth, it may contain petroleum hydrocarbons, or a few other chemicals as well as the natural gas. Most of the petroleum hydrocarbons and other chemicals are removed before the gas is sold to the consumer. The result of this process is a gas that is odorless and colorless. Because natural gas would be hard to detect, an "odorant" is added to it. As a result, the detection of a natural gas leak is made less difficult because of the presence of this chemical. This chemical is added to natural gas as a safety msasure. It's purpose is to give ample warning in the event of a gas leak. Natural gas is (nontoxic) and not poisonous. There is nothing in its composition that is hazardous to your health. The weight of a gas is compared to the weight of air, using air as a standard, a gas is either lighter or heavier than air. Gas is classified by a specific gravity, the ratio of the density to that of air. Air has a specific gravity of 1.0. A gas with a specific gravity of 1.5 would be heavier than air, and a gas with a specific gravity of 0.4 would be lighter than air. Natural gas has a specific gravity of about 0.65 and is lighter than air. A gas that is lighter than air is a good safety factor, if there is a gas leak, the gas will rise into the atmosphere and will dissipate. Some fuels, such as propane and butane, are heavier than air and will settle to the ground, such as a basement. Natural gas is a clean fuel, and will burn almost completely with little or no residue. Natural gas must be mixed with air to burn. As with any type of fire, oxygen is necessary to support combustion. Gas will burn only when it is mixed with certain proportions of air, if the air and gas are not correct the gas will not burn. The air-gas proportions have definite limits called the "upper flammable limit" and the "lower flammable limit." A gas will burn only when the air-gas mix is within these limits. The upper flammable limit is a mixture of 14% gas and 86% air, if the amount of gas exceeds 14%, the mixture will be too rich and will not burn. The lower flammable limit is a gas-air mixture containing only 4% gas and 96% air. Less than 4% gas will be too lean and will not burn. To ignite a correctly air-gas mixture must be heated to a temperature of 1,100 °F. The flame of a match exceeds 1,100 °F and will ignite the mixture. Whenever natural gas burns, the result of its combustion is, of course, heat. When natural gas burns *completely,* two products are released along with the heat, H_2O and CO_2. When natural gas is not burned completely, you will have H_2O, CO_2 and CO carbon monoxide, and is toxic. Some of the early gas burners were open ends of pipe. The flame from this type of burner was a bright yellow. This type of burner gave off light as well as heat. Better burners were developed. The open gas pipe was discarded for a burner with many openings or ports this burner became known as the multiport burner. The flame from these burners were yellow and luminous, so the burners were called luminous burners. Looking at the flame from such a burner shows a small blue colored area at the port or opening followed by a much larger area that is bright, luminous, and yellow in color. Natural gas is a hydrocarbon (hydrogen and carbon). When the gas is ignited, the hydrogen burns at a greater speed than the carbon and burns with a bluish color. The bright area of the flame is the burning carbon. Hydrogen burns at a greater speed because it reacts with the oxygen in the air quicker than the carbon does. The slower burning carbon forms into semisolid particles which are heated to incandescence and make the luminous and yellow in color. Carbon particles complete their combustion when they receive the necessary amount of air. The air around the flame provides the oxygen. The carbon will receive the necessary oxygen for combustion when it reaches the outer surface of the flame, combustion is completed only on the outer surface of the flame. The inside of the flame contains carbon, it will not burn because there is no oxygen to support combustion. The oxygen necessary to support the combustion of a luminous or yellow flame is taken from the air surrounding the flame. When a gas flame does not receive a sufficient air, the gas will not be completely burned. Incomplete combustion, the products are toxic, and are aldehydes (toxic products acrid in odor and irritating to the eyes, nose and throat), and carbon monoxide, these products are harmful and result from incomplete combustion. Incomplete combustion also occurs when a yellow flame is cooled below its ignition temperature, by touching a cold surface. The carbon particles in the cooled area of the flame do not burn completely. These particles are called soot. Soot is released because the partially burned carbon particles in the flame become cold by the cooler surface. They fail to continue burning and therefore adhere to the surface. The products of incomplete combustion from a yellow flame are:

1. Soot
2. Aldehydes
3. Water vapor
4. Carbon monoxide
5. Carbon dioxide

A yellow flame needs a large area to burn, the burners on a gas range are not the luminous type.

In 1855 Robert Wilhelm Von Bunsen discovered a new an entirely different combustion process, when he mixed air with the gas before burning it. So significant was his discovery that the burner was named in his honor, and was called the Bunsen burner. The flame from the Bunsen burner is altogether different from that of the yellow flame burner. It's blue in color, and smaller in size, and higher in temperature. The flame produced by the Bunsen burner consists of several parts, the most prominent being the inner bright blue cone. Around this cone is an outer mantle, or shroud much less brilliant but also blue in color.

In a laboratory type Bunsen burner, gas and air enter at the bottom of the tube. The flame will be at the upper end of the barrel. The air entering to mix with the gas is the primary air.

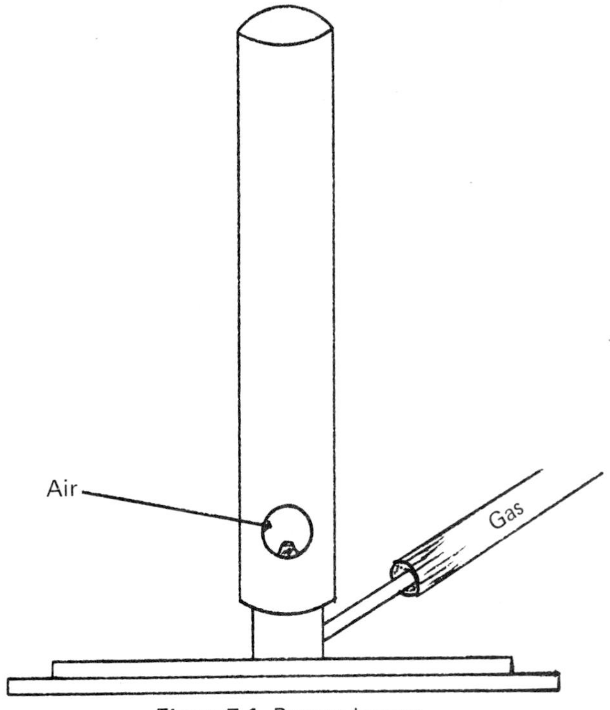

Figure 7.1. Bunsen burner.

In a conventional Bunsen burner approximately 50% of the air necessary for combustion is mixed with the gas before it is ignited. The remainder of the air necessary for combustion, is called secondary air, is supplied from around the flame. Combustion in the Bunsen burner flame occurs in two zones. Oxygen from the primary air first combines with the gas and the initial reaction takes place on the very thin surface of the inner cone. A rapid chemical reaction takes place in this bright blue area. Oxygen in the secondary air mixes with the products of combustion causing final and complete combustion to take place in the outer mantle, or shroud. Surrounding the flame is a barely visible envelope of very hot combustion products. Should any burnable gases still be present, they will be burned in this region. There are no semi-solid carbon particles in a blue flame because there is enough oxygen present in the primary air to react with all of the carbon in the gas, so the carbon remains in a gaseous state throughout the combustion process. If there is too little primary air to a Bunsen burner, there will not be enough oxygen available to react with all of the carbon, some of the carbon will be formed into semi-solid particles. The carbon particles that formed because the amount of primary air was not sufficient will become incandescent and appear as a yellow tip on the cone of the blue flame. A complete restriction of primary air in a Bunsen burner will result in a flame such as that from the yellow flame burner. The addition of primary air speeds up the combustion of the gas flame and the resulting flame is smaller in size. Lack of primary air in a gas flame causes a thermal breakdown of the hydrocarbons of the gas into carbon and hydrogen. It's the breaking down process that slows the combustion time of the yellow flame. The blue flame is more concentrated, its temperature is higher than that of the yellow flame. If the primary air to the burner is excessive, the flame will be distorted and will lift away the burner port. It will make a sharp blowing noise and could blow itself out. Incomplete combustion can result from a blowing flame, with a subsequent release of carbon monoxide. Even though there is sufficient primary air is supplied to the burner, without sufficient secondary the burner will have a yellow flame. Blue flames are adaptable to many applications because of their size and character. This is the reason why blue flames are widely used on modern gas appliances.

Operation of the Honeywell V800 Gas Valve

Figures 7.4 and 7.5 are schematic diagrams which illustrate the operation of the main valve which controls the gas to the burner, the on-off cycling function of the valve operator, and the servo pressure regulator method of maintaining the desired outlet pressure to the burner manifold. The basic theory is applicable to all models in the V800 combination gas controls.

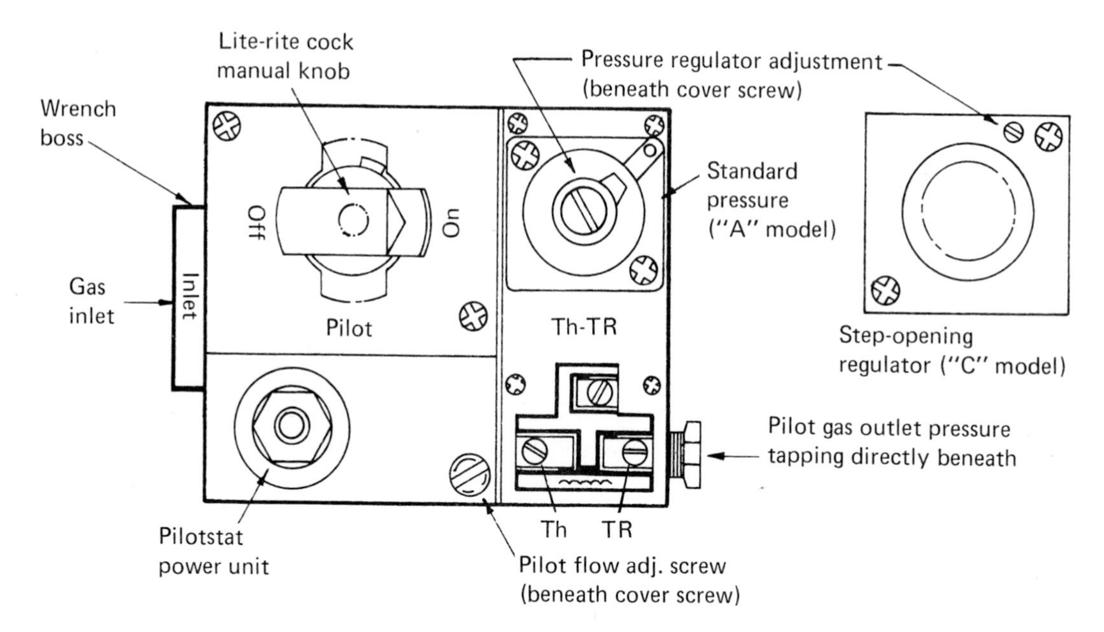

Figure 7.2. Top view of standard capacity model V800.

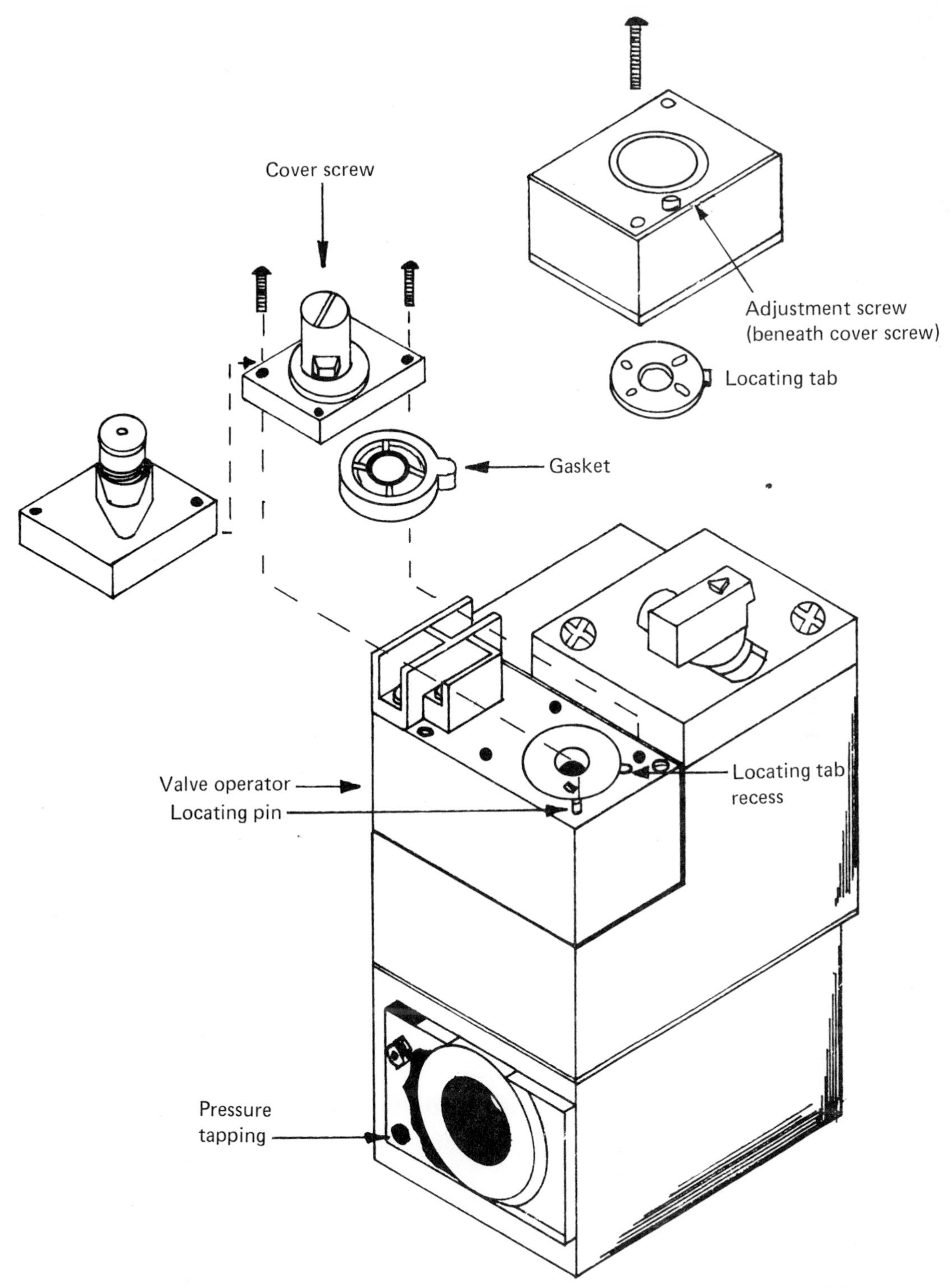

Figure 7.3. Installing regulator or combination gas control with value operator V800.

Burner Off Condition—Figure 7.4

The main valve assembly, Diaphragm and Disc, operates similar to a conventional diaphragm type valve. The valve opens and closes in response to admission or discharge of gas into the pressure chamber beneath the diaphragm. In Figure 7.4 the gas has been discharged such that there is no lifting force being exerted by the diaphragm on the valve disc assembly. The valve closing spring has firmly seated the valve disc blocking the flow of gas to the burner.

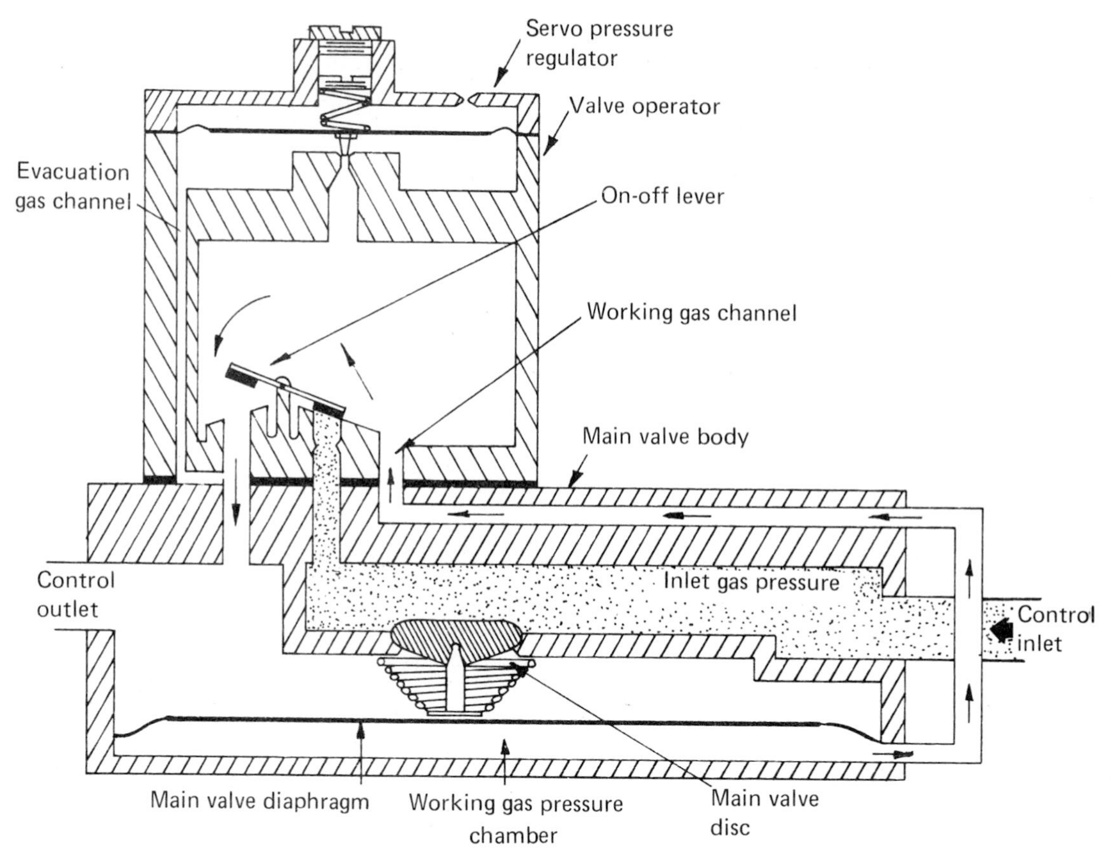

Figure 7.4. V800 gas valve in the burner off position.

Valve Operator

The function of the valve operator is to control the flow of gas. It incorporates an On-Off lever which is electrically actuated by the temperature control circuit. In Figure 7.4 the burner off position, note that the right-hand port is closed, blocking the flow of gas into the gas channel from the control inlet. Also, the evacuation gas channel to the control outlet is open. This has permitted discharge of working gas from the pressure chamber, and with the loss of this gas pressure, the main valve has closed.

Burner on Position Figure 7.5

In Figure 7.5 the operator On-Off lever is in the position it assumes when the valve operator is energized on a call for heat. The left-hand outlet port is closed, and the right-hand supply port is open, permitting the flow of gas into the main valve pressure chamber. The admission of this gas into this chamber causes upward movement of the diaphragm and lifting of the valve disc to allow flow of gas to the burner.

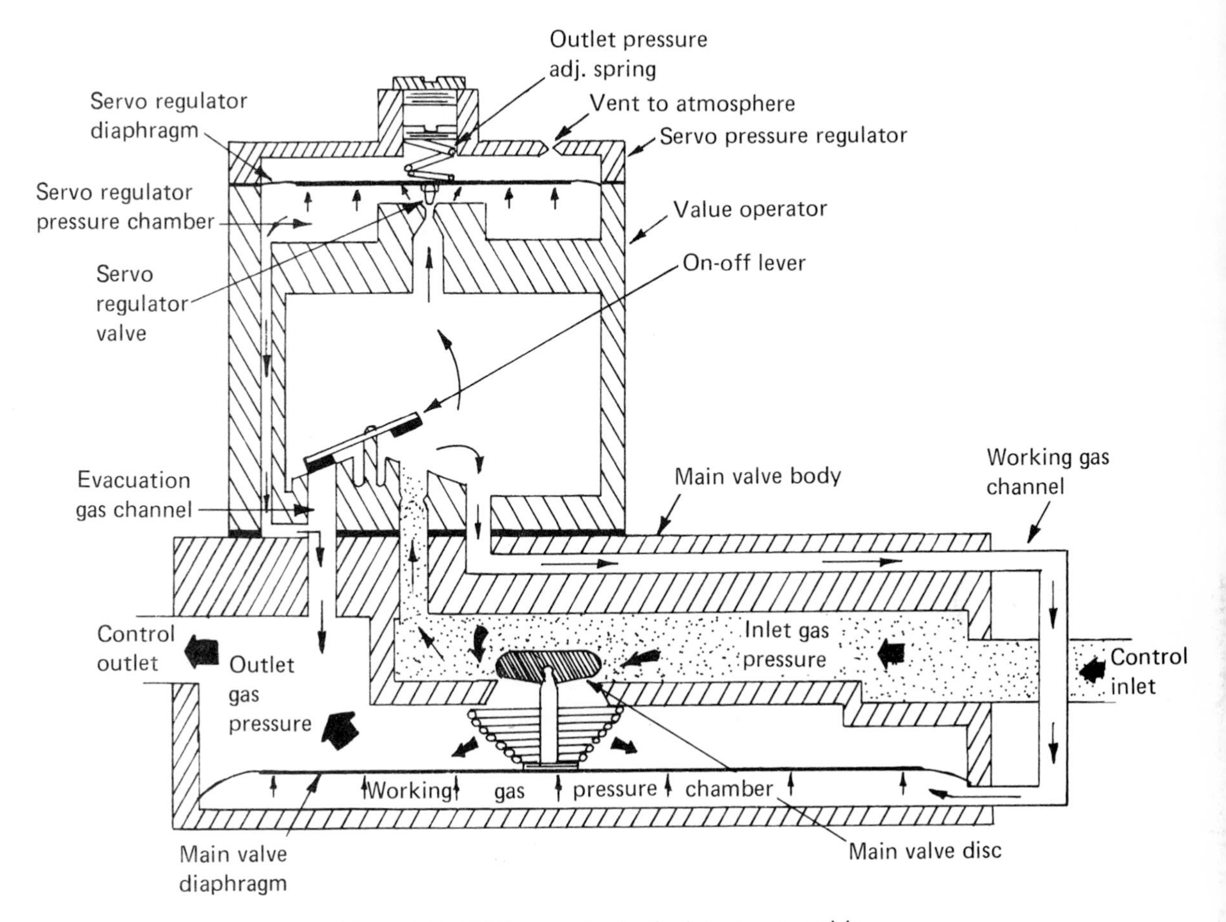

Figure 7.5. V800 gas valve in the burner on position.

Servo Pressure Regulation Figure 7.5

When the right-hand supply port in the valve operator opens upon a call for heat, a continuous flow of gas also is established into the servo regulator pressure chamber, and through the evacuation gas channel into the control outlet. It is through the evacuation channel that the servo regulator senses variation in outlet pressure. Any such variation in outlet pressure is instantly reflected back into the servo regulator pressure chamber and repositions the regulator diaphragm. Movement of the diaphragm in turn alters the rate of flow of evacuation gas through the regulator valve, and cause the following corrective action: If outlet pressure begins to rise, the servo regulator valve opens slightly to allow more gas to discharge into the evacuation gas channel. This decreases the pressure of the working gas in the main valve pressure chamber and repositions the main valve disc downward, closer to its seat. The flow of main burner gas through the control thus is diminished to correct the rise in outlet pressure. If outlet pressure begins to fall, the servo regulator valve closes slightly to reduce the discharge of working gas into the evacuation gas channel.

This increases the pressure of the working gas in the main valve pressure chamber and repositions the main valve disc upward, further away from its seat. The flow of main burner gas through the control thus is increased to correct the fall in outlet pressure.

Safety Shutoff by Pilotstat Mechanism Figure 7.6

Figure 7.6 illustrates the pilotstat safety shutoff, which causes shutdown if the pilot flame is extinguished or becomes too small for satisfactory burner ignition. The mechanism is shown in the normal operating position, Figure 7.6. The pilot is burning, constantly heating the thermocouple (or Powerpile generator) to provide the electrical energy required by the pilotstat power unit to "hold-in" against the force of the compressed loading (drop out) spring. The safety shutoff valve disc at the left end of the rocker arm is in the lower, open position allowing the supply of gas to enter the interior of the tapered plug gas cock.

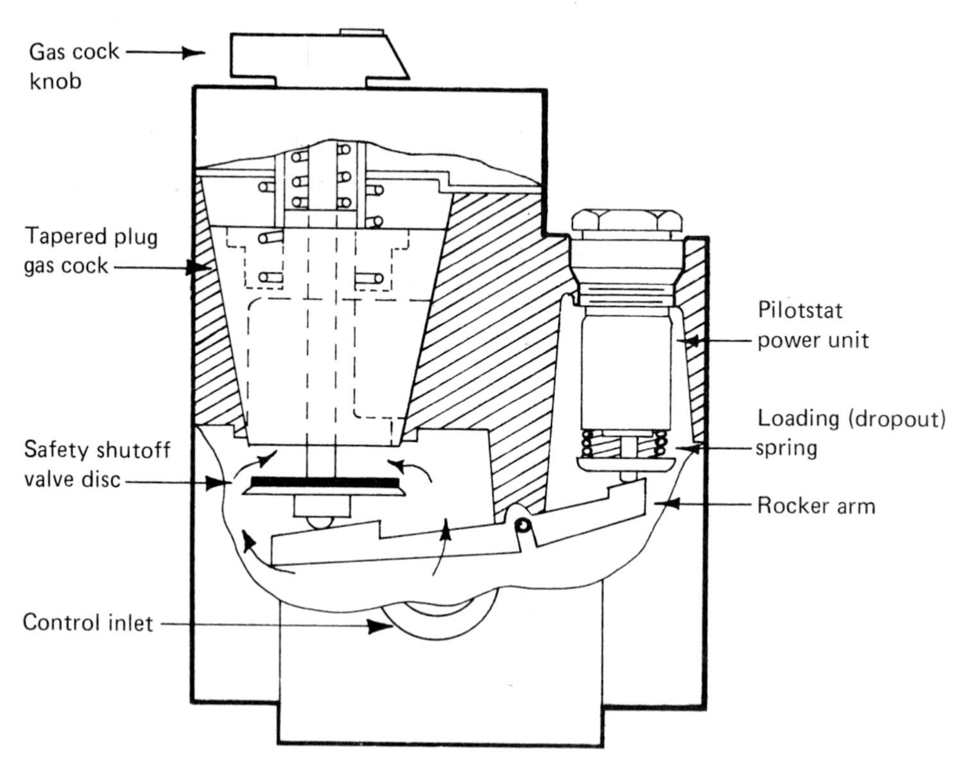

Figure 7.6. Cutaway view of pilotstat mechanism in A, normal operating position.

With the gas cock in the **On** position, supply gas is admitted to the inlet gas pressure chamber illustrated in Figure 7.5. On a call for heat, the main valve disc opens and allows gas to flow to the main burner as described in burner on position. In the event of pilot outage (or decrease in thermocoupls or generator output due to a flame condition), the loading spring in Figure 7.6 causes the power unit to "drop out"—seating the safety shutoff valve disc against the bottom entrance port of the gas cock. Gas flow to both main burner and pilot is physically blocked. It is necessary to relight the pilot and manually reset the pilotstat mechanism to resume operation.

Liquefied Petroleum Gas

Liquefied petroleum gas (L.P.G.) is derived from petroleum by distillation. When petroleum is distilled, heat is applied to the point where the petroleum boils. This heat is not sufficient to cause combustion. However, it is sufficient to cause the gases to be released. They are then trapped and compressed under pressure where they return to a liquid. For efficient transportation and use, L.P. gases are handled in cylinders or steel bottles. The two main L.P. gases are propane and butane. Butane (C_4H_{10}) has a boiling point of +32 °F. Therefore, it cannot be used where the temperature drops below that temperature. Propane (C_3H_8) is used more extensively than butane as its boiling point is −44 °F. Propane has a heating value of 2,500 BTU's per cubic foot of gas at a standard temperature and pressure and a theoretical flame temperature of 3,660 °F.

Most propane gas systems are single stage. That is, one regulator does the entire job of reducing tank or cylinder pressure down to 11 in. of water column, appliance pressure in one step or stage. This may mean reducing as high as 200 lbs. down to 11 in. in warm weather (see Fig. 7.8). When you go into two stage regulation, you divide the pressure reduction into two steps or stages. Two regulators are used and the job of reducing the pressure is split between

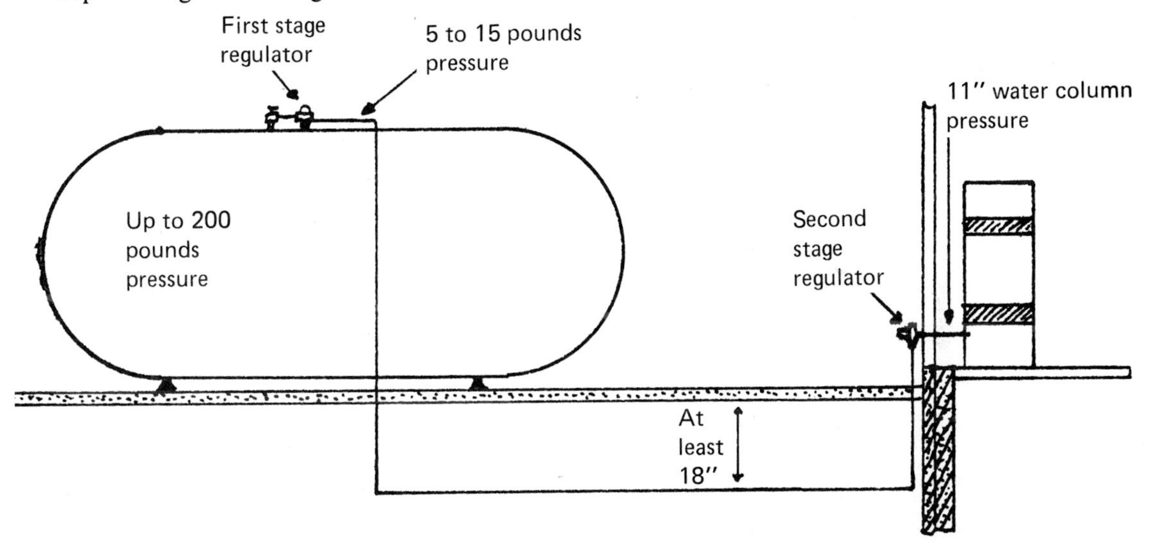

Figure 7.7. Two stage pressure regulation.

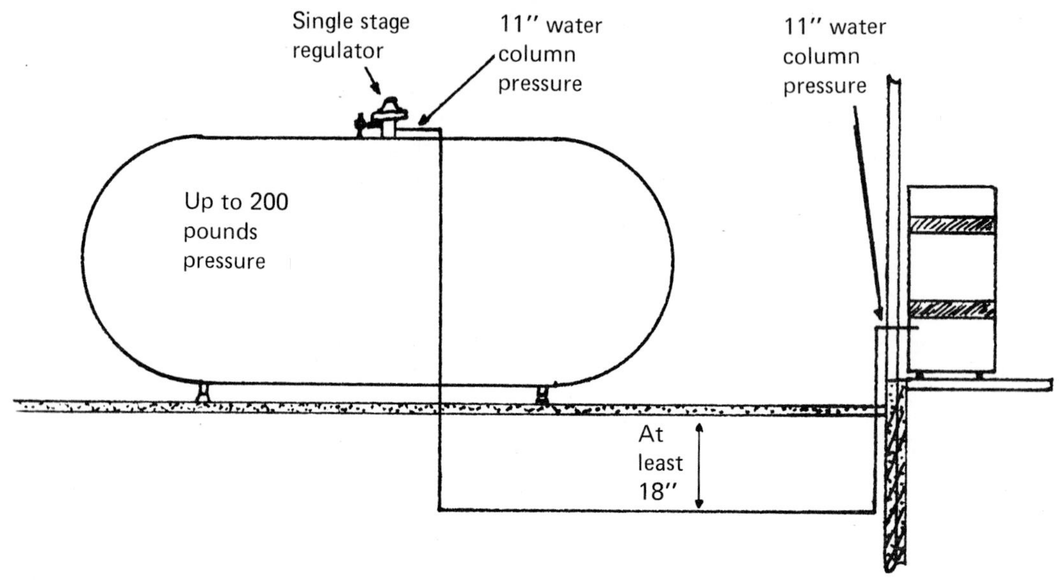

Figure 7.8. Single pressure regulation.

them. The first regulator (first stage) reduces tank or cylinder pressure down to an intermediate pressure ranging from 5 to 15 lbs. This regulator is located at the tank or cylinder. The second regulator (second stage) receives this intermediate pressure and reduces it down to 11 in. water column appliance pressure. This regulator is located just outside the house or building where the gas line goes inside (see Fig. 7.7).

In a single stage system, even a few inches of water column pressure drop in the line to the building is not good for appliance operation. In a two stage system, even a 3 or 4 lb. pressure drop doesn't matter. This is because the pressure drop occurs in the intermediate pressure ahead of the second stage. The second stage regulator receives this intermediate pressure and delivers very close to 11 in. water column to the house piping regardless of substantial pressure drops from the tank to the house.

If excessive pressure drop is to be avoided, it is expensive to run piping to several different outlets. If the piping is to operate at 11 in. water column, such a job might be at a tourist court. It is less expensive to install a first stage regulator at the tank and then install a second stage regulator at each point of use. The pipe in between can be smaller and the appliances will operate better because of two staging. A two stage system will often continue to operate without freezing up even when single stage systems fill with ice. A regulator never freezes—it is the moisture in the fuel which freezes. The size pipe used between the first and second stage regulators isn't too critical, as a 3 to 4 lb. variation in intermediate pressure is taken care of by the second stage regulator. If your first stage regulator is set at 5 to 10 pounds (this is the intermediate pressure); a 15 lbs. setting will use the same table without any trouble. Perhaps your load size came out to be 50 cu. ft. per hr. and your line from the tank to the building is 75 ft. long. Table 7.1 says use 3/8 in. O.D. copper tubing between regulators. If you had a single stage system, the pipe size would have to be 1 1/4 in. in order to allow for the inevitable drop in the piping inside the building.

Table 7.1
If Your Line Between Regulators (tank to building) Is This Long

		25 feet	50 feet	75 feet	100 feet
Use this size tubing or pipe to keep pressure drop below 2 lbs. For the maximum flow shown.	50 CFH 125,000 B.T.U./Hr.	3/8″ O.D. Copper Tubing	3/8″ O.D. Copper Tubing	3/8″ O.D. Copper Tubing	3/8″ O.D. Copper Tubing
	100 CFH 250,000 B.T.U./Hr.	3/8″ O.D. Copper Tubing	3/8″ O.D. Copper Tubing	3/8″ O.D. Copper Tubing	1/2″ O.D. Copper Tubing
	150 CFH 375,000 B.T.U./Hr.	1/2″ O.D. Copper Tubing	1/2″ O.D. Copper Tubing	1/2″ O.D. Copper Tubing	1/2″ O.D. Copper Tubing
	200 CFH 500,000 B.T.U./Hr.	1/2″ O.D. Copper Tubing	1/2″ O.D. Copper Tubing	1/2″ O.D. Copper Tubing	1/2″ O.D. Copper Tubing
	300 CFH 750,000 B.T.U./Hr.	1/2″ O.D. Copper Tubing	3/4″ Std. pipe	3/4″ Std. pipe	3/4″ Std. pipe

Table 7.2
If Your Line from the Second Stage Regulator to the Appliance Is This Long Based on 11 in. W.C.

		10 feet	20 feet	30 feet	40 feet	50 feet
Use this size tubing or pipe to keep pressure drop below 1/2″ water column for the maximum flow shown	10 CFH 25,000 B.T.U./Hr.	3/8″ O.D. Copper Tubing	3/8″ O.D. Copper Tubing	1/2″ O.D. Copper Tubing	1/2″ O.D. Copper Tubing	1/2″ O.D. Copper Tubing
	20 CFH 50,000 B.T.U./Hr.	1/2″ O.D. Copper Tubing	1/2″ O.D. Copper Tubing	1/2″ O.D. Copper Tubing	5/8″ O.D. Copper Tubing	5/8″ O.D. Copper Tubing
	30 CFH 75,000 B.T.U./Hr.	1/2″ O.D. Copper Tubing	5/8″ O.D. Copper Tubing	5/8″ O.D. Copper Tubing	5/8″ O.D. Copper Tubing	5/8″ O.D. Copper Tubing
	50 CFH 125,000 B.T.U./Hr.	5/8″ O.D. Copper Tubing	5/8″ O.D. Copper Tubing	3/4″ Pipe	3/4″ Pipe	3/4″ Pipe
	75 CFH 187,500 B.T.U./Hr.	3/4″ Pipe	3/4″ Pipe	3/4″ Pipe	3/4″ Pipe	3/4″ Pipe

Table 7.3

If This Is Your Lowest Outdoor Temperature (average for 24 hour period)

		32°F	20°F	10°F	0°F	−10°F	−20°F	−30°F
Gas need to vapor-ize per hour. (Not absolute maximum, but average rate of withdrawal in 8 hour period.) Tank at least this big. And keep half full.	50-CFH 125,000 B.T.U./Hr.	115 Gal.	115 Gal.	115 Gal.	250 Gal.	250 Gal.	400 Gal.	600 Gal.
	100-CFH 250,000 B.T.U./Hr.	250 Gal.	250 Gal.	250 Gal.	400 Gal.	500 Gal.	1000 Gal.	1500 Gal.
	150-CFH 375,000 B.T.U./Hr.	300 Gal.	400 Gal.	500 Gal.	500 Gal.	1000 Gal.	1500 Gal.	2500 Gal.
	200-CFH 500,000 B.T.U./Hr.	400 Gal.	500 Gal.	750 Gal.	1000 Gal.	1200 Gal.	2000 Gal.	3500 Gal.
	300-CFH 750,000 B.T.U./Hr.	750 Gal.	1000 Gal.	1500 Gal.	2000 Gal.	2500 Gal.	4000 Gal.	5000 Gal.

Troubleshooting Gas Burners

Pilot Go Out When Reset Knob Is Released

1. Reset knob or button released too soon. Must hold in for 60 sec.
2. Thermocouple or thermopile is bad.
3. Pilot flame not enveloping 3/8″ to 1/2″ of thermocouple or thermopile.
4. Thermocouple or thermopile terminals shorted or loose.
5. Connection dirty or wet.
6. Pilotstat power unit is bad.
7. Pilot filter clogged.

Pilot Cannot Be Lighted

1. Gas supply is turned off.
2. Gas supply not purged of air.
3. Pilot orifice is clogged.
4. Lighting knob not set at pilot position.
5. Reset button not being depressed.
6. Lighting knob not set at the pilot position.
7. Pilot gas flow adjustment is closed off.

Pilot Outage During System Use

1. Pilot orifice is clogged.
2. Thermocouple or thermopile is bad.

3. Pilotstat power unit is bad.
4. Gas supply pressure too low.
5. Thermocouple cold junction or too hot.
6. Connection dirty, loose or wet.
7. Pilot flame not enveloping 3/8" to 1/2".
8. Pilot burner dirt screen clogged.
9. Pilot filter clogged.
10. Gas meter vent clogged.
11. Pilot unshielded from excessive draft.
12. Pilot may go out when blower comes on. Look for cracked heat exchanger.

Pilot Flame Yellow or Lazy

1. Pilot burner primary air opening clogged.
2. Pilot orifice too large.
3. Pilot lint screen clogged.
4. Clogged appliance venting or incorrect.

Pilot Flame Noisy, Blowing or Lifting

1. Pilot gas pressure too high.
2. Pilot line has air in it.

Pilot Flame Hard or Sharp

1. Pilot orifice too small.
2. Incorrect pilot orifice.
3. Typical of butane and propane pilots.

Pilot Flame Blue and Small

1. Pilot orifice clogged.
2. Pilot gas flow adjustment closed off.
3. Pilot orifice too small.
4. Pilot filter clogged.
5. Pilot line clogged.

Pilot Flame Blue and Waving

1. Gas pressure too low.
2. Pilot line clogged.
3. Pilot unshielded from products of combustion.

Troubleshooting Thermostats

Thermostat Jumpered System Won't Work

1. Thermostat not at fault, check elsewhere.
2. Limit control set too low.
3. Low voltage control circuit open.
4. Low voltage transformer burned out.

5. Gas valve is bad.
6. Bad terminals or wire.

Thermostat Jumpered System Works

1. Thermostat contacts are dirty.
2. Thermostat damaged.

Room Temp Overshoots Thermostat Setting

1. Thermostat mounted on cold wall.
2. Heating unit too large.
3. Thermostat not properly calibrated.
4. Thermostat not mounted level (mercury switch type).
5. Thermostat wiring hole not plugged.
6. Thermostat exposed to cold drafts.
7. Thermostat heater set too high.
8. Thermostat not exposed to circulating air.
9. Thermostat does not have a heater.

Room Temp Doesn't Reach Thermostat Setting

1. Thermostat not mounted level (mercury switch type).
2. Thermostat not properly calibrated.
3. Heating unit too small.
4. Limit control set too low.
5. Thermostat exposed to direct rays of sun.
6. Thermostat affected by heat from fireplace.
7. Thermostat affected by lamp or appliances.
8. Thermostat affected by stove or oven.
9. Thermostat mounted on warm wall.
10. Thermostat mounted near register or radiator.
11. Thermostat contacts are dirty.
12. Bad terminals.
13. Clogged filter in forced air system.

Thermostat Seems Out of Calibration

1. Thermostat not mounted level (mercury switch type).
2. Thermostat not properly calibrated.
3. Bad terminals.
4. Clogged air filter.

Thermostat Cycles too Often

1. Thermostat heater set too low.

Thermostat Doesn't Cycle Often Enough

1. Thermostat not exposed to circulating air.
2. Heating unit too small.
3. Thermostat set too high.
4. Thermostat does not have a heater.
5. Thermostat contacts are dirty.

Room Temp Swings Excessively

1. Thermostat not exposed to circulating air.
2. Thermostat heater set too high.
3. Heating plant too large.
4. Thermostat does not have a heater.

Sec. For One Rev.	Meter Dial Used–CU Ft. Per Rev.		Sec. For One Rev.	Meter Dial Used–CU Ft. Per Rev.			Sec. For One Rev.	Meter Dial Used–CU Ft. Per Rev.		
	1/2	1		1/2	1	2		1	2	5
10	180	360	35	52	103	206	60	60	120	300
11	164	327	36	50	100	200	62	58	116	290
12	150	300	37	49	97	195	64	56	112	281
13	139	277	38	48	95	189	66	55	109	273
14	129	257	39	46	92	185	68	53	106	265
15	120	240	40	45	90	180	70	52	103	257
16	113	225	41	44	88	176	72	50	100	250
17	106	212	42	43	86	172	74	49	97	243
18	100	200	43	42	84	167	76	48	95	237
19	95	189	44	41	82	164	78	46	92	231
20	90	180	45	40	80	160	80	45	90	225
21	86	171	46	39	78	157	84	43	86	214
22	82	164	47	38	77	153	88	41	82	205
23	78	157	48	38	75	150	92	39	78	196
24	75	150	49	37	74	147	96	38	75	188
25	72	144	50	36	72	144	100	36	72	180
26	69	138	51	35	71	141	105	34	69	172
27	67	133	52	35	69	138	110	33	66	164
28	64	129	53	34	68	136	120	30	60	150
29	62	124	54	33	67	133	130	28	55	138
30	60	120	55	33	66	131	140	26	52	129
31	58	116	56	32	64	129	150	24	48	120
32	56	113	57	32	63	126	160	23	45	113
33	55	109	58	31	62	124	170	21	43	106
34	53	106	59	31	61	122	180	20	40	100

Clocking method for checking gas input. To convert meter flow rate (CFH) to B.T.U. per hour, multiply CFH by the B.T.U. heat content of the gas being used.

Drill Size Chart for Main Burner Orifices

	Natural Gas 1,020 BTU-.65 SG 3-1/2" WC Manifold		Propane 2,500 BTU-1.5 SG 11" WC Manifold	
Input BTU/hr per Spud	Drill Size	Decimal Tolerance	Drill Size	Decimal Tolerance
12,000 15,000	51 48	.064-.007 .073-.076	60 58	.038-.040 .040-.042
20,000 25,000	43 41	.086-.089 .093-.096	55 53	.050-.052 .056-.059
30,000 40,000	39 32	.097-.100 .113-.116	51 49	.064-.067 .070-.073
50,000 60,000	30 27	.124-.128 .140-.144	46 43	.078-.081 .086-.089
70,000 80,000	22 20	.153-.157 .156-.161	42 40	.090-.093 .095-.098
90,000 100,000	17 13	.168-.173 .180-.185	38 35	.098-.101 .107-.110
105,000 110,000	11 10	.186-.191 .188-.193	34 33	.108-.111 .109-.113
125,000 135,000	5 3	.200-.205 .208-.213	1/8 30	.121-.125 .124-.128
140,000 150,000	7/32 1	.214-.219 .233-.228	30 29	.124-.128 .132-.136
160,000 175,000	A C	.229-.234 .237-.242	28 27	.136-.140 .140-.144
190,000 200,000	E F	.245-.250 .252-.257	25 23	.145-.149 .150-.154
210,000 220,000	H I	.261-.266 .267-.272	21 20	.154-.159 .156-.161
240,000 260,000	K M	.276-.281 .290-.295	18 16	.164-.169 .172-.177
280,000 300,000	5/16 O	.307-.312 .311-.316	13 11	.180-.185 .186-.191
310,000 320,000	P 21/64	.318-.323 .323-.328	9 7	.191-.196 .196-.201

Orifice Diameter Inches	Pilot gas consumption in BTU/hr for type of gas and orifice pressure indicated.		
	Natural Gas at 3.5″ W.C.	Natural Gas at 7″ W.C.	Propane at 11″ W.C.
.009	200	280	560
.010	250	350	710
.011	300	425	860
.012	355	500	1,010
.013	430	600	1,210
.014	480	690	1,400
.015	565	800	1,620
.016	640	900	1,840
.018	785	1,120	2,270
.020	980	1,390	2,810
.021	1,070	1,500	3,040
.0225	1,225	1,730	
.024	1,390	1,960	
.025	1,530	2,160	
.026	1,660	2,350	
.028	1,920	2,720	
.030	2,210	3,120	

Natural Gas–1,000 BTU–Cu. Ft. 0.65 SP. GR.
Propane Gas–2,500 BTU–Cu. Ft. 1.53 Sp. GR.

Installation of Gas Appliances and Gas Piping*

by the American Gas Association

1.1.1 *Applicability:*

This standard applies to the design, fabrication, installation, tests and operation of appliance and piping systems for fuel gases such as natural gas, manufactured gas, undiluted liquefied petroleum gases, liquefied petroleum gas-air mixtures, or mixtures of any of these gases as follows:

A. *Low pressure* (not in excess of 1/2 pound per square inch or 14 inches water column) domestic and commercial piping systems extending from the outlet of the meter set assembly, or the outlet of the service regulator when a meter is not provided, to the inlet connections of appliances.

B. The installation and operation of domestic and commercial appliances supplied at pressures of 1/2 pound per square inch or less.

*Available from the American Standards Association Inc., 10 East 40th Street, New York, New York 10016, or the American Society of Mechanical Engineers, 345 East 47th Street, New York, New York 10017.

1.1.2 *Nonapplicability*

This standard does not apply to:

A. Gas piping systems for industrial installations at any pressure or any other gas piping system operating at pressures greater than 1/2 pound per square inch. For piping in such installations refer to A.S.M.E. code for pressure piping, Section 2 of A.S.A. B31.1-1955 and Addenda B31.1A-1961*.

B. Gas equipment designed and installed for specific manufacturing, production, processing and power generating applications.

C. Gas equipment supplied through piping systems covered in 1.1.2(A).

1.1.3 *Other Standards:*

In applying this standard, reference should also be made to the manufacturer's instructions, servicing gas supplier regulations, and local building, heating, plumbing or other codes in effect in the area in which the installation is made.

1.1.4 *Approved:*

The word "approved," as used in this standard, means acceptable to the authority having jurisdiction.

1.2 *Qualified Installing Agency*

Installation and replacement of gas piping or gas appliances and repair of gas appliances shall be performed only by a qualified installing agency. By the term "Qualified Installing Agency" is meant any individual, firm, corporation, or company which either in person or through a representative is engaged in and is responsible for the installation or replacement of gas piping on the outlet side of the meter, or of the service regulator when a meter is not provided, or the connection, installation or repair of gas appliances, who is experienced in such work, familiar with all precautions required, and has complied with all the requirements of the authority having jurisdiction.

1.3 *General Precautions*

1.3.1 *Turn Gas Off:*

All gas piping or gas appliance installation shall be performed with the gas turned off to eliminate hazards from leakage of gas.

1.3.2 *Notification of Interrupted Service:*

It shall be the duty of the installing agency when the gas supply is to be turned off, to notify all affected consumers.

1.3.3 *Before Turning Gas Off:*

Before turning off the gas to premises for the purpose of installation, repair, replacement or maintenance of gas piping or appliances, all burners shall be turned off. When two or more consumers are served from the same supply system, precautions shall be exercised to assure that only service to the proper consumer is turned off.

1.3.4 *Checking for Gas Leaks:*

Soap and water solution, or other material acceptable for the purpose shall be used in locating gas leakage. Matches, candles, flame or other sources of ignition shall not be used for this purpose.

1.3.5 *Use of Lights:*

Artificial illumination used in connection with a search for gas leakage shall be restricted to electric hand flashlights (preferably of the safety type) or approved safety lamps. In searching for leaks, electric switches should not be operated. If electric lights are already turned on, they should not be turned off.

1.3.6 *Working Alone:*

An individual shall not work alone in any situation where accepted working practice dictates that two or more men are necessary to perform the work safely.

1.3.7 *Handling of Liquid from Drips:*

Liquid which is removed from a drip in existing gas piping shall be handled with proper precautions, and shall not be left on the consumer's premises.

1.3.8 *No Smoking:*

When working on piping which contains or has contained gas, smoking shall not be permitted.

1.3.9 *Handling Flammable Liquids:*

Flammable liquids used by the installer shall be handled with proper precautions and shall not be left within the premises from the end of one working day to the beginning of the next.

1.3.10 *Work Interruptions:*

When interruptions in work occur, the system shall be left in a safe and satisfactory condition.

Gas Piping Installation

2.1 *Piping Plan:*

It is recommended that before proceeding with the installation of a gas piping system, a piping sketch or plan be prepared showing the proposed location of the piping as well as the size of different branches. Adequate consideration should be given to future demands and provisions made for added gas service. Before any final plans or specifications are completed, the serving gas supplier or the authority having jurisdiction should be consulted when an additional appliance is to be served through any present gas piping, capacity of the existing piping shall be checked for adequacy, and replaced with larger piping if necessary.

2.2 *Provision for Meter Location:*

The meter location shall be such that the meter can be easily read and the connections are readily accessible for servicing. Location, space requirements, dimensions and type of installation shall be acceptable to the serving gas supplier. Gas piping at multiple meter installations shall be plainly marked by a metal tag or other permanent means attached by the installing agency, designating the building or the part of the building being supplied.

2.3 *Interconnections:*

2.3.1 Interconnections supplying separate consumers:
When two or more meters, or two or more service regulators when meters are not provided, are installed on the same premises and supply separate consumers, the gas piping systems shall not be interconnected on the outlet side of the meters or service regulators.

2.3.2 *Interconnections for Stand-by Fuels:*
When a supplementary gas for stand-by use is connected down-stream from a meter or a service regulator when a meter is not provided, a suitable device to prevent backflow shall be installed. A three-way valve installed to admit the stand-by supply and at the same time shut off the regular supply may be used for this purpose.

2.4 *Size of Piping to Gas Appliances:*

2.4.1 Size of Supply Piping for Gas Appliances:
Gas piping shall be of such size and so installed as to provide a supply of gas sufficient to meet the maximum demand without undue loss of pressure between the meter or service regulator when a meter is not provided and the appliance or appliances. The size of gas piping depends upon the following factors:

A. Allowable loss in pressure from meter, or service regulator when a meter is not provided, to appliance.
B. Maximum gas consumption to be provided.
C. Length of piping and number of fittings.
D. Specific gravity of the gas.
E. Diversity factor.

2.4.2 *Gas Consumption:*
The quantity of gas to be provided at each outlet shall be determined, whenever possible, directly from the manufacturer's BTU rating of the appliance which will be installed. In case the ratings of the appliances to be installed are not known, Table 1 is given to show the approximate consumption of average appliances of certain types in BTU per hour. To obtain the cubic feet per hour of gas required, divide the total BTU input of all appliances by the average BTU heating value per cubic foot of the gas. The average BTU per cubic foot of the gas in the area of the installation may be obtained from the serving gas supplier.

2.4.3 *Gas Piping Size:*

A. Capacities in cubic feet per hour of 0.60 specific gravity gas for different sizes and length are shown in tables 2A and 2B for iron pipe or equivalent rigid pipe and in table 2C for semi-rigid tubing. Tables 2A and 2C are based upon a pressure drop of 0.3 inch water column, whereas table 2B is based upon a pressure drop of 0.5 inch water column. In using these tables no additional allowance is necessary for an ordinary number of fittings. The serving gas supplier shall designate which table(s) shall be used.
B. Capacities in thousands of BTU per hour of undiluted liquefied petroleum gases based on a pressure drop of 0.5 inch water column for different sizes and lengths

are shown in table 4A for iron pipe or equivalent rigid pipe and in table 4B for semi-rigid tubing. In using these tables no additional allowance is necessary for an ordinary number of fittings.

C. Gas piping systems that are to be supplied with gas of a specific gravity of 0.70 or less, can be sized directly from tables 2A, 2B and 2C unless the authority having jurisdiction specifies that a gravity factor be applied. When the specific gravity of the gas is greater than 0.70 the gravity factor shall be applied. Application of the gravity factor converts the figures given in tables 2A, 2B and 2C to capacities with another gas of different specific gravity. Such application is accomplished by multiplying the capacities given in tables 2A, 2B and 2C by the multipliers shown in table 3. In case the exact specific gravity does not appear in the table, choose the next higher value specific gravity shown.

D. To determine the size of each section of gas piping in a system within the range of tables 2A, 2B, 2C, 4A or 4B proceed as follows:

1. Determine the gas demand of each appliance to be attached to the piping system. When tables 2A, 2B, or 2C are to be used to select the piping size, calculate the gas demand in terms of cubic feet per hour for each piping system outlet. When tables 4A or 4B are to be used to select the piping size, calculate the gas demand in terms of thousands of BTU per hour for each piping system outlet.
2. Measure the length of piping from the gas meter or service regulator when a meter is not provided, to the most remote outlet in the building.
3. In tables 2A, 2B, 2C, 4A or 4B, which ever is appropriate, select the column showing that distance or the next longer distance if the table does not give the exact length. This is the only distance used in determining the size of any section of gas piping. If the gravity factor is to be applied, the values in the selected column of tables 2A, 2B or 2C are multiplied by the appropriate multiplier from table 3.
4. Use this vertical column to locate all gas demand figures for this particular system of piping.
5. Starting at the most remote outlet, find in the vertical column just selected, the gas demand for the outlet. If the exact figure of demand is not shown, choose the next larger figure below in the column.
6. Opposite this demand figure, in the first column at the left in tables 2A, 2B, 2C, 4A or 4B, will be found the correct size of gas piping.
7. Proceed in a similar manner for each outlet and each section of gas piping. For each section of piping determine the total gas demand supplied by the section.

E. For any gas piping system, for special gas appliances or for conditions other than those covered by tables 2A, 2B, 2C, 4A or 4B, such as longer runs, or greater gas demands, the size of each gas piping system shall be determined by standard engineering methods acceptable to the authority having jurisdiction and the serving gas supplier.

2.4.4 *Diversity Factor:*

The diversity factor is an important factor in determining the correct gas piping size to be used in multiple family dwellings. It is dependent upon the number and kinds of gas appliances being installed. Consult the serving gas supplier or the authority having jurisdiction for the diversity factor to be used.

2.4.5 *Additions to Existing Gas Piping:*

Additions to existing utility gas piping shall conform to tables 2A, 2B, or 2C, whichever is designated by the serving gas supplier. Additions to existing undiluted liquefied petroleum gas piping shall conform to table 4A or 4B. Existing gas piping that does not conform to these provisions shall be replaced by the proper size of pipe or tubing. Additions shall not be made to existing pipe which is smaller than that

Table 1

Approximate Gas Input for Some Common Appliances

Appliance	Input BTU per hr. (Approx.)
Range, free standing domestic	65,000
Built-in oven or broiler unit, domestic	25,000
Built-in top unit, domestic	40,000
Water heater, automatic storage 30 to 40 gal. tank	45,000
Water heater, automatic storage 50 gal. tank	55,000
Water heater, automatic instantaneous	
Capacity (2 gal. per minute)	142,800
Capacity (4 gal. per minute)	285,000
Capacity (6 gal. per minute)	428,400
Water heater, domestic, circulating or side-arm	35,000
Refrigerator	3,000
Clothes dryer, type 1 (domestic)	35,000
Gas light	2,500
Incinerator, domestic	35,000
For specific appliances or appliances not shown above, the input should be determined from the manufacturer's rating.	

Table 2A

Maximum Capacity of Pipe in Cubic Feet of Gas per Hour

(Based upon a pressure drop of 0.3 inch water column and 0.6 specific gravity gas)

Length in Feet	Nominal Iron Pipe Size, Inches								
	1/2	3/4	1	11/4	11/2	2	21/2	3	4
10	132	278	520	1,050	1,600	3,050	4,800	8,500	17,500
20	92	190	350	730	1,100	2,100	3,300	5,900	12,000
30	73	152	285	590	890	1,650	2,700	4,700	9,700
40	63	130	245	500	760	1,450	2,300	4,100	8,300
50	56	115	215	440	670	1,270	2,000	3,600	7,400
60	50	105	195	400	610	1,150	1,850	3,250	6,800
70	46	96	180	370	560	1,050	1,700	3,000	6,200
80	43	90	170	350	530	990	1,600	2,800	5,800
90	40	84	160	320	490	930	1,500	2,600	5,400
100	38	79	150	305	460	870	1,400	2,500	5,100
125	34	72	130	275	410	780	1,250	2,200	4,500
150	31	64	120	250	380	710	1,130	2,000	4,100
175	28	59	110	225	350	650	1,050	1,850	3,800
200	26	55	100	210	320	610	980	1,700	3,500

Table 2B

Maximum Capacity of Pipe in Cubic Feet of Gas per Hour

(Based upon a pressure drop of 0.5 inch water column and 0.6 specific gravity gas)

Length in Feet	Nominal Iron Pipe Size, Inches								
	1/2	3/4	1	11/4	11/2	2	2 1/2	3	4
10	175	360	680	1,400	2,100	3,950	6,300	11,000	23,000
20	120	250	465	950	1,460	2,750	4,350	7,700	15,800
30	97	200	375	770	1,180	2,200	3,520	6,250	12,800
40	82	170	320	660	990	1,900	3,000	5,300	10,900
50	73	151	285	580	900	1,680	2,650	4,750	9,700
60	66	138	260	530	810	1,520	2,400	4,300	8,800
70	61	125	240	490	750	1,400	2,250	3,900	8,100
80	57	118	220	460	690	1,300	2,050	3,700	7,500
90	53	110	205	430	650	1,220	1,950	3,450	7,200
100	50	103	195	400	620	1,150	1,850	3,250	6,700
125	44	93	175	360	550	1,020	1,650	2,950	6,000
150	40	84	160	325	500	950	1,500	2,650	5,500
175	37	77	145	300	460	850	1,370	2,450	5,000
200	35	72	135	280	430	800	1,280	2,280	4,600

Table 2C
Maximum Capacity of Semirigid Tubing in Cubic Feet of Gas per Hour
(Based on a pressure drop of 0.3 inch water column and
0.6 specific gravity gas)

Outside Diameter (Inches)	Length of Tubing (Feet)									
	10	**20**	**30**	**40**	**50**	**60**	**70**	**80**	**90**	**100**
3/8	19	12	10	9	—	—	—	—	—	—
1/2	45	30	24	20	18	17	15	14	13	12
5/8	97	64	52	44	38	35	32	30	28	26
3/4	161	105	88	71	64	59	54	50	46	44
7/8	245	169	135	114	97	91	80	75	71	67

Table 3
Multipliers to Be Used Only with Tables 2A, 2B and 2C
When Applying the Gravity Factor

Specific Gravity	Multiplier	Specific Gravity	Multiplier
.35	1.31	1.00	.78
.40	1.23	1.10	.74
.45	1.16	1.20	.71
.50	1.10	1.30	.68
.55	1.04	1.40	.66
.60	1.00	1.50	.63
.65	.96	1.60	.61
.70	.93	1.70	.59
.75	.90	1.80	.58
.80	.87	1.90	.56
.85	.84	2.00	.55
.90	.82	2.10	.54

permitted in tables 2A, 2B or 4A, or to existing tubing which is smaller than that permitted in table 2C or 4B.

2.5 *Gas Piping in Mobile Home and Travel Trailer Parks:*
Gas piping systems in mobile home and travel trailer parks extending from the outlet of a meter set assembly or the outlet of a service regulator when a meter is not provided to the terminal of the gas riser at each trailer site shall comply with the following specific provisions and with all other applicable provisions in part 1 and part 2 of this standard.

2.5.1 *Protection of Piping:*
Piping shall be buried to a sufficient depth or covered in a manner so as to protect the piping system from physical damage.

Table 4A

Maximum Capacity of Pipe in Thousands of BTU per Hour
of Undiluted Liquefied Petroleum Gases
(Based on a pressure drop of 0.5 inch water column)

Nominal Iron Pipe Size, Inches	10	20	30	40	50	60	70	80	90	100	125	150
1/2	275	189	152	129	114	103	96	89	83	78	69	63
3/4	567	393	315	267	237	217	196	185	173	162	146	132
1	1,071	732	590	504	448	409	378	346	322	307	275	252
11/4	2,205	1,496	1,212	1,039	913	834	771	724	677	630	567	511
11/2	3,307	2,299	1,858	1,559	1,417	1,275	1,181	1,086	1,023	976	866	787
2	6,221	4,331	3,465	2,992	2,646	2,394	2,205	2,047	1,921	1,811	1,606	1,496

Table 4B

Maximum Capacity of Semirigid Tubing in Thousands of BTU
per Hour of Undiluted Liquefied Petroleum Gases
(Based on a pressure drop of 0.5 inch water column)

Outside Diameter (Inches)	10	20	30	40	50	60	70	80	90	100
3/8	39	26	21	19	—	—	—	—	—	—
1/2	92	62	50	41	37	35	31	29	27	26
5/8	199	131	107	90	79	72	67	62	59	55
3/4	329	216	181	145	131	121	112	104	95	90
7/8	501	346	277	233	198	187	164	155	146	138

2.5.2 *Prohibited Locations:*
Piping shall not be installed under trailer sites and patio slabs adjacent to trailers when an enclosing foundation is used under the trailer.

2.5.3 *Location, Protection and Sizing of Riser:*
The gas riser to each trailer site should be placed in the rear one-third section of the site and not less than 18 inches from the roadside wall of the trailer. It shall be located and protected or supported so as to minimize the liklihood of damage by moving vehicles. The minimum size of the gas piping outlet at a trailer site shall be 3/4 inch for other than undiluted liquefied petroleum gases.

2.5.4 *Location of Shutoff Valves:*
(a) Outlets for the individual trailers and gas piping to any building supplied by the system shall be provided with a readily accessible approved valve which cannot be locked in the open position.

(b) A readily accessible valve shall be provided near the point of gas delivery for shutting off the entire trailer park system. The valve provided by the serving gas supplier may be considered acceptable for this purpose provided it is readily accessible.

2.5.5 *Connection of Trailer:*

Trailers shall be connected to the gas piping system with rigid pipe, listed connectors or semirigid tubing. Connectors having aluminum exterior surfaces shall not be used.

2.5.6 *Demand Factors:*

(a) The hourly volume of gas required for any trailer site, gas outlet or any section of a trailer park gas piping system may be computed from Table 5.

(b) Other gas equipment or appliances, other than trailer site outlets, shall be computed at the manufacturer's maximum cubic foot per hour input rating or from Table 1 and shall be added to the figures given in Table 5.

Table 5

Demand Factors for Use in Calculating Gas Piping Systems in Trailer Parks

No. of Trailer Sites	BTU per Hour per Trailer Site
1	125,000
2	117,000
3	104,000
4	96,000
5	92,000
6	87,000
7	83,000
8	81,000
9	79,000
10	77,000
11-20	66,000
21-30	62,000
31-40	58,000
41-60	55,000
over 60	50,000

2.6 *Acceptable Piping Materials*

2.6.1 *Piping Material:*

(a) Pipe. Gas pipe shall be steel or wrought-iron pipe complying with the American standard for wrought-steel and wrought-iron pipe, ASA B.36 10-1959*. Threaded copper, brass, or aluminum alloy pipe in iron pipe sizes may be used with gases not corrosive to such material except that aluminum alloy pipe shall not be used in exterior

*Available from the American Standards Association, Inc., 10 East 40th Street, New York, New York 10016.

locations, or underground, or where it is in contact with masonry, plaster, or insulation, or is subject to repeated corrosive wettings. Aluminum alloy pipe shall comply with specification ASTM B-241 (except that the use of alloy 5456 is prohibited) and shall be suitably marked at each end of each length indicating compliance with ASTM Specifications.**

(b) Tubing. When acceptable to the serving gas supplier, seamless copper aluminum alloy, or steel tubing may be used with gases not corrosive to such material. Copper tubing shall be of standard type K or L, or equivalent, complying with specification ASTM B88-62 and having a minimum wall thickness for each tubing size in compliance with ASTM specifications.* Aluminum alloy tubing shall be of standard Type A or B, or equivalent, complying with specification ASTM B-318-62, having a minimum wall thickness for each tubing size, and being suitably marked every 18 inches in compliance with ASTM specifications.* Aluminum alloy tubing shall not be used in exterior locations, or underground, or where it is in contact with masonry, plaster, or insulation, or is subject to repeated corrosive wettings.

(c) Piping joints and fittings. Pipe joints may be screwed, flanged, or welded, and nonferrous pipe may also be soldered or brazed with material having a melting point in excess of 1,000 °F. Tubing joints shall either be made with approved flared gas tubing fittings, or be soldered or brazed with a material having a melting point in excess of 1,000 °F. Compression type tubing fittings shall not be used for this purpose.

Fittings (except stockcocks or valves) shall be malleable iron or steel when used with steel or wrought-iron pipe and shall be copper or brass when used with copper or brass pipe or tubing, and shall be aluminum alloy when used with aluminum alloypipe or tubing. When approved by the authority having jurisdiction, special fittings may be used to connect steel or wrought-iron pipe. Cast-iron fittings in sizes 6 inches and larger may be used to connect steel and wrought-iron pipe when approved by the authority having jurisdiction.

2.6.2 *Workmanship and Defects:*

Gas pipe or tubing and fittings shall be clear and free from cutting burrs and defects in structure or threading and shall be thoroughly brushed, and chip and scale blown. Defects in pipe or tubing or fittings shall not be repaired. When defected pipe, tubing or fittings are located in a system the defective material shall be replaced.

2.6.3 *Pipe Coating:*

When in contact with material exerting a corrosive action, piping and fittings coated with a corrosion resisting material shall be used.

2.6.4 *Use of Old Piping Material:*

Gas pipe, tubing, fittings and valves removed from any existing installation shall not be again used until they have been thoroughly cleaned, inspected and ascertained to be equivalent to new material.

*Available from American Society for Testings and Materials, 1916 Race St., Philadelphia, Pa. 19103.

2.6.5 *Joint Compounds:*
Joint compounds (pipe dope) shall be applied sparingly and only to the male threads of pipe joints. Such compounds shall be resistant to the action of liquefied petroleum gases.

2.7 *Pipe Threads*

2.7.1 *Specifications for Pipe Threads:*
Pipe and fitting threads shall comply with the American standard for pipe threads (except dryseal), B2.1-1960*.

2.7.2 *Damaged Threads:*
Pipe with threads which are stripped, chipped, corroded, or otherwise damaged shall not be used.

2.7.3 *Number of Threads:*
Pipe shall be threaded in accordance with Table 6.

Table 6
Specifications for Threading Pipe

Iron Pipe Size (Inches)	Approximate Length of Threaded Portion (Inches)	Approximate No. of Threads to Be Cut
1/2	3/4	10
3/4	3/4	10
1	7/8	10
1 1/4	1	11
1 1/2	1	11
2	1	11
2 1/2	1 1/2	12
3	1 1/2	12
4	1 5/8	13

2.8 *Concealed Piping in Buildings*

2.8.1 *Minimum Size:*
No gas pipe smaller than standard 1/2 inch iron pipe size shall be used in any concealed location.

2.8.2 *Piping in Partitions:*
Concealed gas piping should be located in hollow rather than solid partitions. Tubing shall not be run inside walls or partitions unless protected against physical damage. This rule does not apply to tubing which passes through walls or partitions.

*Available from the American Standards Association, Inc., 10 East 40th Street, New York, New York 10016.

2.8.3 *Piping in Floors:*

(a) Except as provided in 2.8.3 (b), gas piping in solid floors such as concrete shall be laid in channels in the floor suitably covered to permit access to the piping with a minimum of damage to the building. When piping in floor channels may be exposed to excessive moisture or corrosive substances, it shall be suitably protected. (b) When approved by the authority having jurisdiction and acceptable to the serving gas supplier, gas piping may be embedded in concrete floor slabs constructed with portland cement. Piping shall be surrounded with a minimum of 1 1/2 inches of concrete and shall not be in physical contact with other metallic structures such as reinforcing rods or electrical neutral conductors. When piping may be subject to corrosion at point of entry into concrete slab, it shall be suitably protected from corrosion. Piping shall not be embedded in concrete slabs containing quickset additives or cinder aggregate.

2.8.4 *Connections in Original Installations*

When installing gas piping which is to be concealed, unions, tubing fittings, running threads, right and left couplings, bushings, and swing joints made by combinations of fittings shall not be used.

2.8.5 *Reconnections:*

When necessary to insert fittings in gas pipe which has been installed in a concealed location, the pipe may be reconnected by use of a ground joint union with the nut center-punched to prevent loosening by vibration. Reconnection of tubing in a concealed location is prohibited.

2.9 *Piping Underground*

2.9.1 *Protection of Piping:*

Piping shall be buried a sufficient depth or covered in a manner so as to protect the piping from physical damage.

2.9.2 *Protection Against Corrosion:*

(a) Gas piping in contact with earth or other material which may corrode the piping, shall be protected against corrosion in an approved manner. When dissimilar metals are joined underground, an insulated coupling shall be used. Piping shall not be laid in contact with cinders.

(b) Underground piping for manufactured gas shall be one size larger than that specified by Table 2A or Table 2B, as designated by the serving gas supplier, but in no case less than 1 1/4 inch.

2.9.3 *Piping Through Foundation Wall:*

Underground gas piping, when installed below grade through the outer foundation or basement wall of a building, shall be either encased in a sleeve or otherwise protected against corrosion. The piping or sleeve shall be sealed at the foundation or basement wall to prevent entry of gas or water.

2.9.4 *Piping Underground Beneath Buildings:*

When the installation of gas piping underground beneath buildings is unavoidable, the piping shall be encased in a conduit. The conduit shall extend into a normally usable and accessible portion of the building and, at the point where the conduit terminates in the building, the space between the conduit and the gas piping shall be sealed to prevent the possible entrance of any gas leakage. The conduit shall extend at least 4 inches outside the building, be vented above grade to the outside and be installed in a way as to prevent the entrance of water.

2.10 *Installation of Piping*

Drips, grading, protection from freezing, and branch pipe connections, as provided for in 2.10.2, 2.10.4, 2.10.7, and 2.10.14(a), shall apply only when other than dry gas is distributed and climatic conditions make such provisions necessary.

2.10.1 *Building Structure:*

The building structure shall not be weakened by the installation of any gas piping. Before any beams or joists are cut or notched, special permission should be obtained from the authority having jurisdiction.

2.10.2 *Gas Piping to be Graded:*

All gas piping shall be graded not less than 1/4 inch in 15 feet to prevent traps. All horizontal lines shall grade to risers and from the risers to the meter, or to service regulator when a meter is not provided, or to the appliance.

2.10.3 *Piping Supports:*

(a) Gas piping in buildings shall be supported with pipe hooks, metal pipe straps, bands or hangers suitable for the size of piping, and of adequate strength and quality, and located at proper intervals so that the piping cannot be moved accidentally from the installed position. Gas piping shall not be supported by other piping.

(b) Spacing of supports in gas piping installations shall not be greater than shown in Table 7.

Table 7

Support of Piping

Size of Pipe Inches	(Feet)	Size of Tubing (Inch O.D.)	(Feet)
1/2	6	1/2	4
3/4 or 1	8	5/8 or 3/4	6
1 1/4 or larger (horizontal)	10	7/8 or 1	8
1 1/4 or larger (vertical)	every floor level		

2.10.4 *Protect Against Freezing:*

Gas piping shall be protected against freezing temperatures. When piping must be exposed to wide ranges or sudden changes in temperatures, special care shall be taken to prevent stoppages.

2.10.5 *Overhanging Rooms:*

When there are overhanging kitchens or other rooms built beyond foundation walls, in which gas appliances are installed, care shall be taken to avoid placing the gas piping where it will be exposed to low temperatures (40 °F or below for manufactured gas) or to extreme changes of temperatures. In such cases the gas piping shall be brought up inside the building proper and run around the sides of the room, in the most practical manner.

2.10.6 *Do Not Bend Pipe:*
Gas pipe shall not be bent. Fittings shall be used when making turns in gas pipe.

2.10.7 *Provide Drips Where Necessary:*
A drip shall be provided at any point in the line of pipe where condensate may collect. When condensate is excessive, a drip shall be provided at the outlet of the meter. This drip should be so installed as to constitute a trap wherein an accumulation of condensate will shut off the flow of gas before it will run back into the meter.

2.10.8 *Location and Size of Drips:*
All drips shall be installed only in such locations that they will be readily accessible to permit cleaning or emptying. A drip shall not be located where the condensate is likely to freeze. The size of any drip used shall be determined by the capacity and the exposure of the gas piping which drains to it and in accordance with recommendations of the serving gas supplier.

2.10.9 *Use Tee:*
If dirt or other foreign material is a problem, a tee fitting with the bottom outlet plugged or capped shall be used at the bottom of any pipe riser. (See Fig. 1.)

2.10.10 *Avoid Clothes Chutes, etc.:*
Gas piping inside any building shall not be run in or through an air duct, clothes chute, chimney or gas vent, ventilating duct, dumb waiter, or elevator shaft.

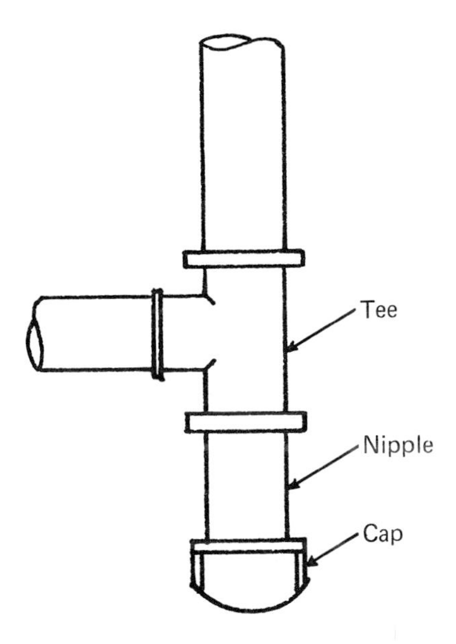

Figure 1. Suggested method of installing tee.

2.10.11 *Cap all Outlets:*

(a) Each outlet, including a valve or cock outlet, shall be securely closed gastight with a threaded plug or cap immediately after installation and shall be left closed until an appliance is connected thereto. Likewise, when an appliance is disconnected from an outlet and the outlet is not to be used again immediately, it shall be securely closed gastight. The outlet shall not be closed with tin caps, wooden plugs, corks, or by other improvised methods.

(b) The above provision does not prohibit the normal use of a listed quick-disconnect device.

2.10.12 *Location of Outlets:*

The unthreaded portion of gas piping outlets shall extend at least one inch through finished ceilings and walls, and when extending through floors shall be not less than 2 inches above them. The outlet fitting or the piping shall be securely fastened. Outlets shall not be placed behind doors. Outlets shall be far enough from floors, walls and ceilings to permit the use of proper wrenches without straining, bending or damaging the piping.

2.10.13 *Prohibited Devices:*

No device shall be placed inside the gas piping or fittings that will reduce the cross-sectional area or otherwise obstruct the free flow of gas.

2.10.14 *Branch Pipe Connection:*

(a) All branch outlet pipes shall be taken from the top or sides of horizontal lines and not from the bottom.

(b) When a branch outlet is placed on a main supply line before it is known what size of pipe will be connected to it, the outlet shall be of the same size as the line which supplies it.

2.10.15 *Electrical Bonding and Grounding:*

(a) A gas piping system within a building shall be electrically continuous and bonded to any grounding electrode, as defined by The National Electrical Code, ASA C1-1962 (NFPA No. 70)*

(b) Underground gas service piping shall not be used as a grounding electrode except when it is electrically continuous uncoated metallic piping, and its use as a grounding electrode is acceptable both to the serving gas supplier and to the authority having jurisdiction, since gas piping systems are often constructed with insulating bushings or joints, or are of coated or nonmetallic piping.

2.11 *Gas Shutoff Valves*

2.11.1 *Accessibility of Gas Valves:*

Main gas shutoff valves controlling several gas piping systems shall be placed an adequate distance from each other so they will be easily accessible for operation and shall be installed so as to be protected from physical damage. It is recommended that they be plainly marked with a metal tag attached by the installing agency so that the gas pip-

*Available from the National Fire Protection Association, 60 Battery March St., Boston, Mass. 02110 in pamphlet form and in the National Fire Codes, Volume 5. Also available from the American Standards Association, Inc., 10 East 40th St., New York, N.Y. 10016.

ing systems supplied through them can be readily identified. It is advisable to place a shutoff valve at every point where safety, convenience of operation, and maintenance demands.

2.11.2 *Shutoff Valves for Multiple House Lines*

A. In multiple tenant buildings supplied through a master meter or one service regulator when a meter is not provided, or where meters or service regulators are not readily accessible from the appliance location, an individual shutoff valve for each apartment, or for each separate house line, shall be provided at a convenient point of general accessibility.

B. In a common system serving a number of individual buildings, shutoff valves shall be installed at each building.

2.12 *Test of Piping for Tightness*

Before any system of gas piping is finally put in service, it shall be carefully tested to assure that it is gas tight. Where any part of the system is to be enclosed or concealed, this test should precede the work of closing in. To test for tightness, the piping may be filled with the fuel gas, air or inert gas, but not with any other gas or liquid. **Oxygen shall never be used.**

A. Before appliances are connected, piping systems shall stand a pressure of at least six inches mercury or three pounds gage for a period of not less than ten minutes without showing any drop in pressure. Pressure shall be measured with a mercury manometer or slope gage, or an equivalent device so calibrated as to be read in increments of not greater than one-tenth pound. The source of pressure shall be isolated before the pressure tests are made.

B. Systems for undiluted liquefied petroleum gases shall stand the pressure test in accordance with 2.12 (A), or, when appliances are connected to the piping system, shall stand a pressure of not less than ten inches water column for a period of not less than ten minutes without showing any drop in pressure. Pressure shall be measured with a water manometer or an equivalent device so calibrated as to be read in increments of not greater than one-tenth inch water column. The source of pressure shall be isolated before the pressure tests are made.

2.13 *Leakage Check After Gas Turn On*

2.13.1 *Close All Gas Outlets:*

Before turning gas under pressure into any piping, all openings from which gas can escape shall be closed.

2.13.2 *Check for Leakage:*

Immediately after turning on the gas, the piping system shall be checked by one of the following methods to ascertain that no gas is escaping:

A. Checking for leakage using the gas meter—Immediately prior to the test it should be determined that the meter is in operating condition and has not been bypassed.

Checking for leakage can be done by carefully watching the test dial of the meter to determine whether gas is passing through the meter. To assist in observing any movement of the test hand, wet a small piece of paper and paste its edge directly over the center line of the hand as soon as the gas is turned on. Allow five minutes for a one-half foot dial and proportionately longer for a larger dial in checking for gas flow.

This observation should be made with the test hand on the upstroke. In case careful observation of the test hand for a sufficient length of time reveals no movement, the piping shall be purged and a small gas burner turned on and lighted and the hand of the test dial again observed. If the dial hand moves (as it should), it will show that the meter is operating properly. If the test hand does not move or register flow of gas through the meter to the small burner, the meter is defective and the gas should be shut off and the serving gas supplier notified.

B. Checking for leakage not using a meter—This can be done by attaching to an appliance orifice a manometer or equivalent device calibrated so that it can be read in increments of 0.1 inch water column, and momentarily turning on the gas supply and observing the gaging device for pressure drop with the gas supply shut off. No discernible drop in pressure shall occur during a period of 3 minutes.

C. When leakage is indicated—If the meter test hand moves, or a pressure drop on the gage is noted, all appliances or outlets supplied through the system shall be examined to see if they are shut off and do not leak. If they are found tight there is a leak in the piping system. The gas supply shall be shut off until the necessary repairs have been made, after which the test specified in 2.13.2 (A) or (B) shall be repeated.

2.14 *Purging*

2.14.1 *Purging All Gas Piping:*

A. After piping has been checked, all gas piping shall be fully purged. A suggested method for purging the gas piping to an appliance is to disconnect the pilot piping at the outlet of the pilot valve. Piping shall not be purged into the combustion chamber of an appliance.

B. The open end of piping systems being purged shall not discharge into confined spaces or areas where there are sources of ignition unless precautions are taken to perform this operation in a safe manner by ventilation of the space, control of purging rate, and elimination of all hazardous conditions.

2.14.2 *Light Pilots:*

After the gas piping has been sufficiently purged, all appliances shall be purged and the pilots lighted. The installing agency shall assure itself that all piping and appliances are fully purged before leaving the premises.

Courtesy of the American Gas Association

Chapter 8
GAS WIRING DIAGRAMS

The following electrical diagrams are typical of gas fired residential equipment.

Figure 8.1a

Gas fired forced air furnace, consisting of

1. Heating Thermostat
2. 24 Volt Gas Valve
3. 30 M.V. Thermocouple
4. 24 Volt Transformer
5. Combination Fan Limit Control
6. Single Speed Blower Motor
7. Electronic Air Cleaner
8. Humidistat
9. 115 Volt Humidifier

Figure 8.1b

Figure 8.1b is the schematic of Figure 8.1a.

Figure 8.2a

Gas fired forced air furnace, consisting of

1. Heating Theomostat
3. 24 Volt Gas Valve
3. 30 M.V. Thermocouple
4. Two 24 Volt Transformer
5. Combination Fan and Limit Control
6. Single Speed blower motor
7. Humidistat
8. 24 Volt Humidifier

Figure 8.2b

Figure 8.2b is the schematic of Figure 8.2a.

Figure 8.3a

Gas fired forced air furnace, prepped for air conditioning, consisting of

1. Heating Cooling Thermostat
2. 24 Volt Gas Valve
3. 30 M.V. Thermocouple
4. 24 Volt Transformer
5. Combination Fan and Limit control
6. Single speed blower motor
7. Humidistat
8. 24 Volt Humidifier
9. Fan Center
10. Electronic Air Cleaner

Figure 8.3b

Figure 8.3b is the schematic of Figure 8.3a

Figure 8.4a

Gas fired forced air furnace prepped for air conditioning consisting of

1. Heating Cooling Thermostat
2. 30 M.V. Thermocouple
3. 24 Volt Gas Valve
4. Combination Fan and Limit Control
5. Multiple speed blower motor
6. Humidistat
7. 115 Volt Humidifier
8. Fan center
9. Electronic Air Cleaner
10. Air flow switch

Figure 8.4b

Figure 8.4b is the schematic of Figure 8.4a.

Figure 8.5a

Gas fired Gravity Furnace, consisting of

1. 750 M.V. Thermopile
2. Limit control
3. Heating Thermostat 750 M.V.
4. Gas Valve 750 M.V.

Figure 8.5b

Figure 8.5b is the scematic of Figure 8.5a.

Figure 8.6a

Gas fired forced air furnace consisting of

1. Gas Valve 750 M.V.
2. 750 M.V. Thermopile
3. Heating thermostat
4. Single speed blower motor
5. Fan Control
6. Limit Control

Figure 8.6b

Figure 8.6b is the schematic of Figure 8.6a.

Figure 8.7a

Gas fired forced air furnace, consisting of

1. Gas Valve 750 M.V.
2. 750 M.V. Thermopile
3. Heating Thermostat
4. Single speed blower motor
5. Combination Fan and Limit Control
6. Humidistat
7. 115 Volt Humidifier

Figure 8.7b

Figure 8.7b is the schematic of Figure 8.7a.

Figure 8.8a

Gas fired forced air furnace consisting of

1. Gas Valve 750 M.V.
2. 750 M.V. Thermopile
3. Heating Cooling Thermostat with heating anticipator turned off
4. Fan Center
5. Combination Fan and Limit Control
6. Humidistat
7. 115 Volt Humidifier
8. Single speed blower motor
9. Electronic Air Cleaner

Figure 8.8b

Figure 8.8b is the schemiatic of Figure 8.8a.

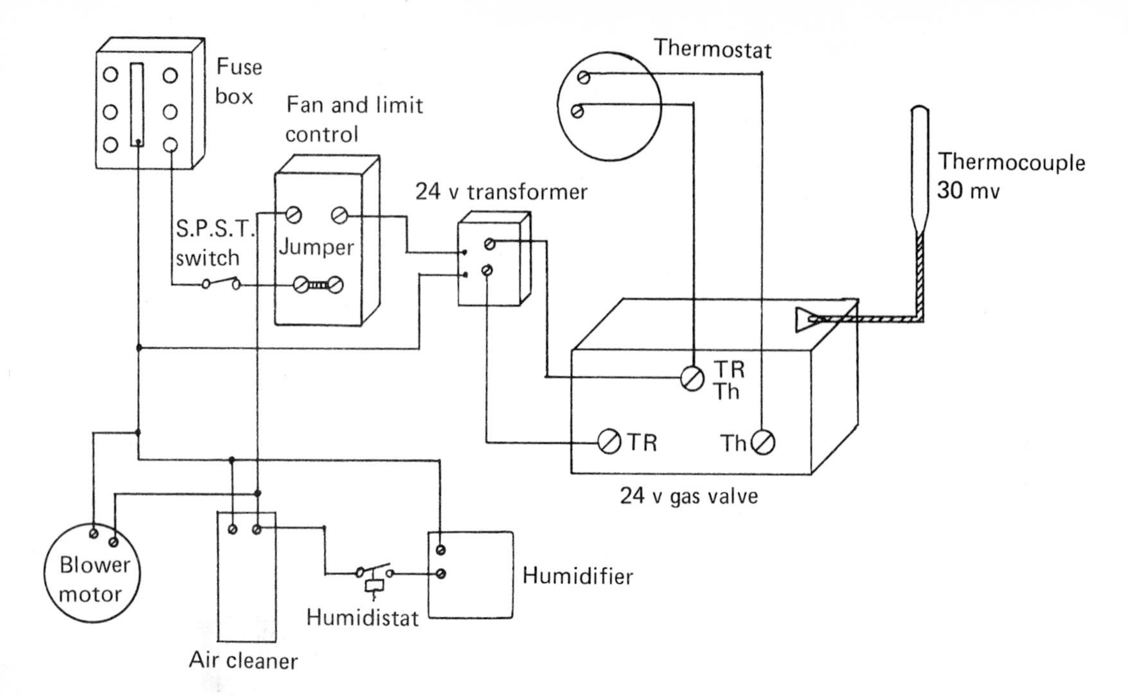

Figure 8.1a.

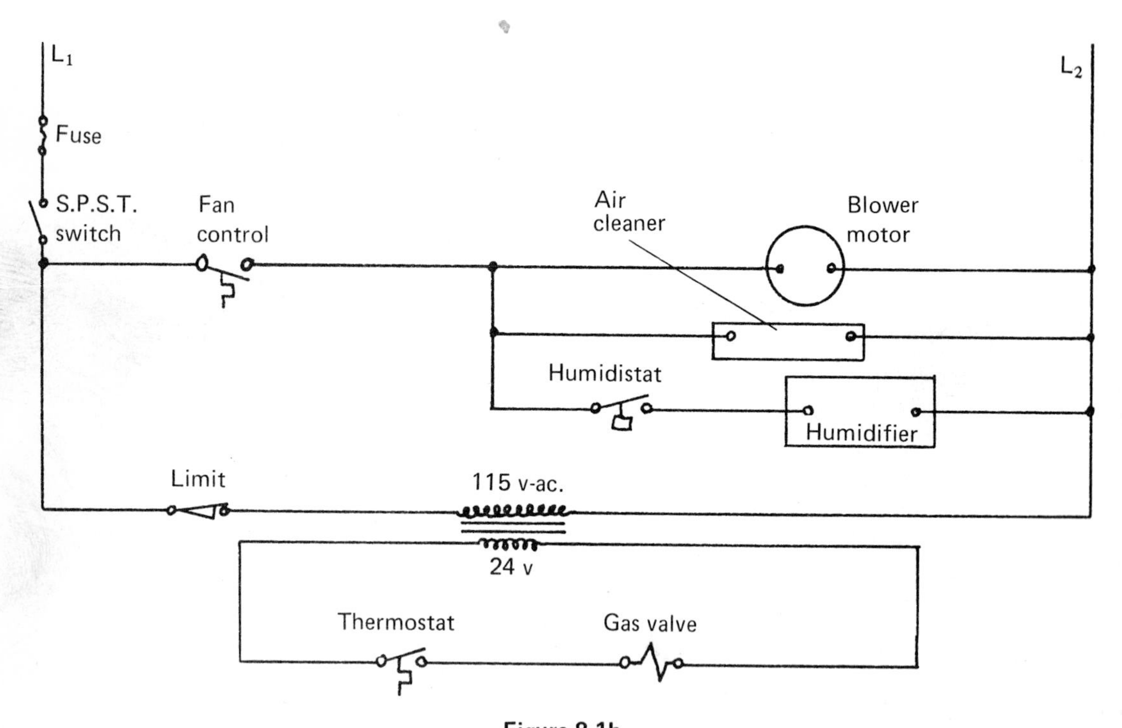

Figure 8.1b.

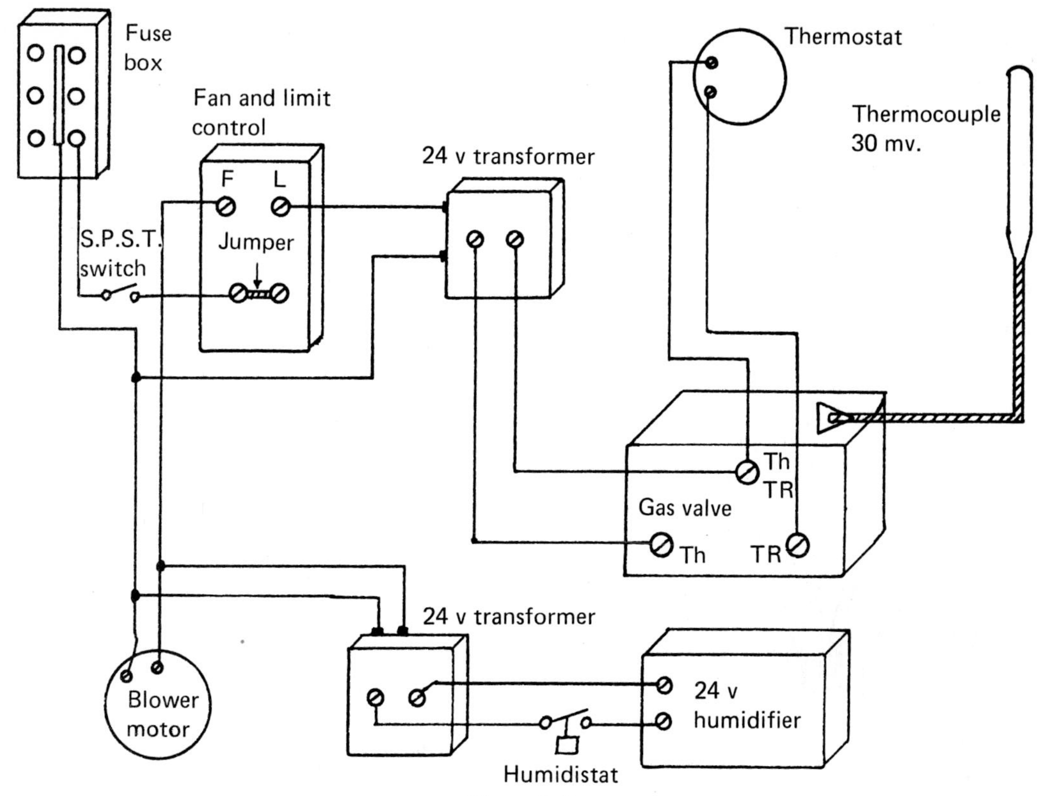

Figure 8.2a.

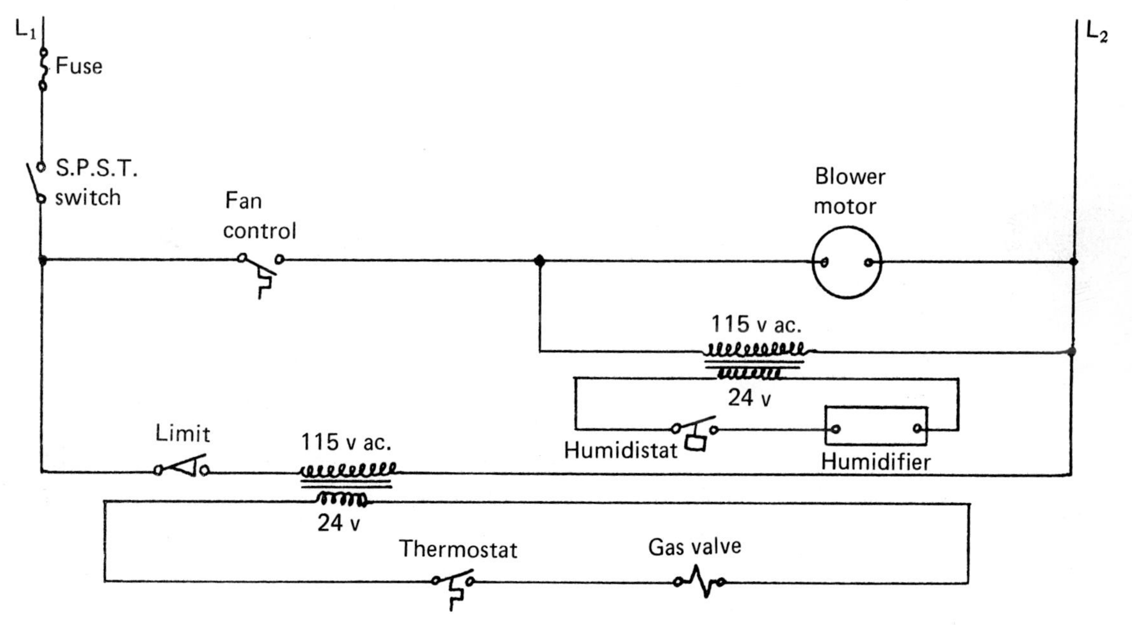

Figure 8.2b.

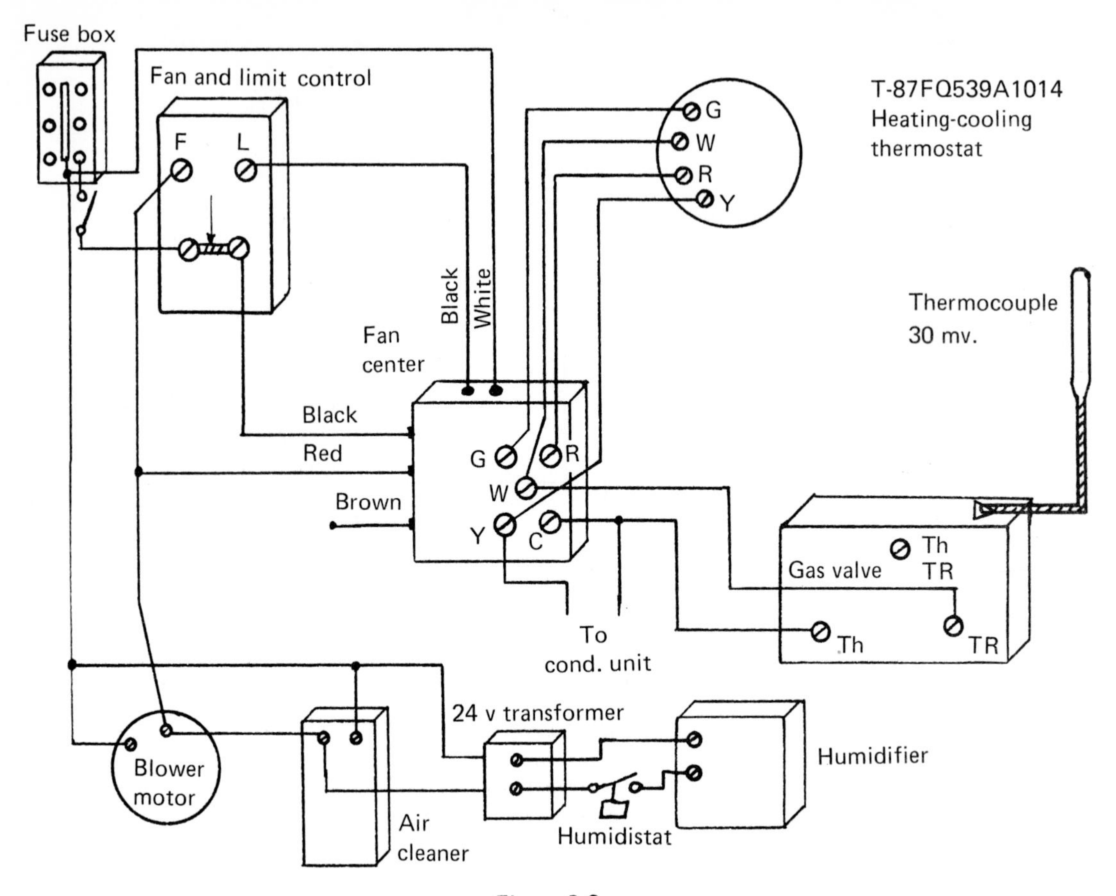

Fuse box
Fan and limit control
F
L
T-87FQ539A1014
Heating-cooling
thermostat
G
W
R
Y
Black
White
Thermocouple
30 mv.
Fan
center
Black
Red
Brown
G
R
W
Y
C
Gas valve
Th
TR
Th
TR
To
cond. unit
24 v transformer
Humidifier
Blower
motor
Air
cleaner
Humidistat

Figure 8.3a.

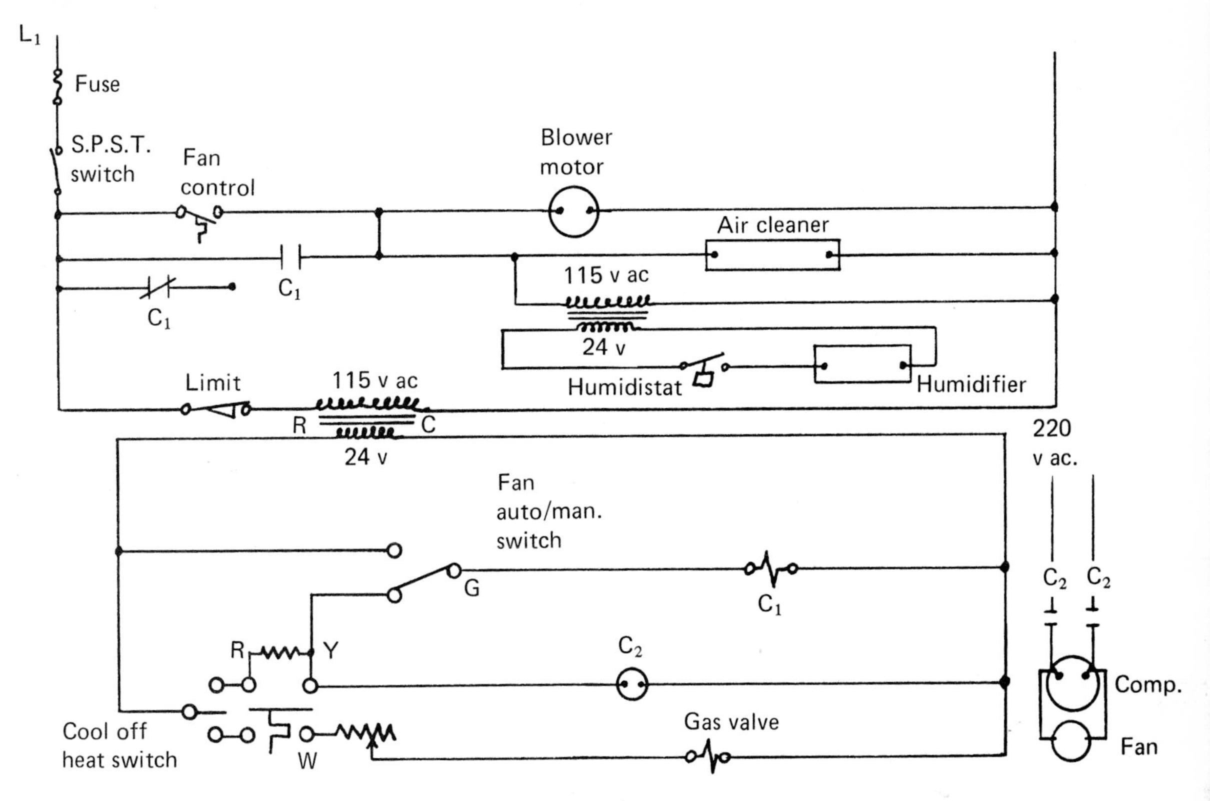

L_1
Fuse
S.P.S.T.
switch
Fan
control
Blower
motor
Air cleaner
C_1
C_1
115 v ac
24 v
Humidistat
Humidifier
Limit
115 v ac
R
C
24 v
220
v ac.
Fan
auto/man.
switch
G
C_1
C_2
C_2
C_2
R
Y
Comp.
Fan
Gas valve
Cool off
heat switch
W

Figure 8.3b.

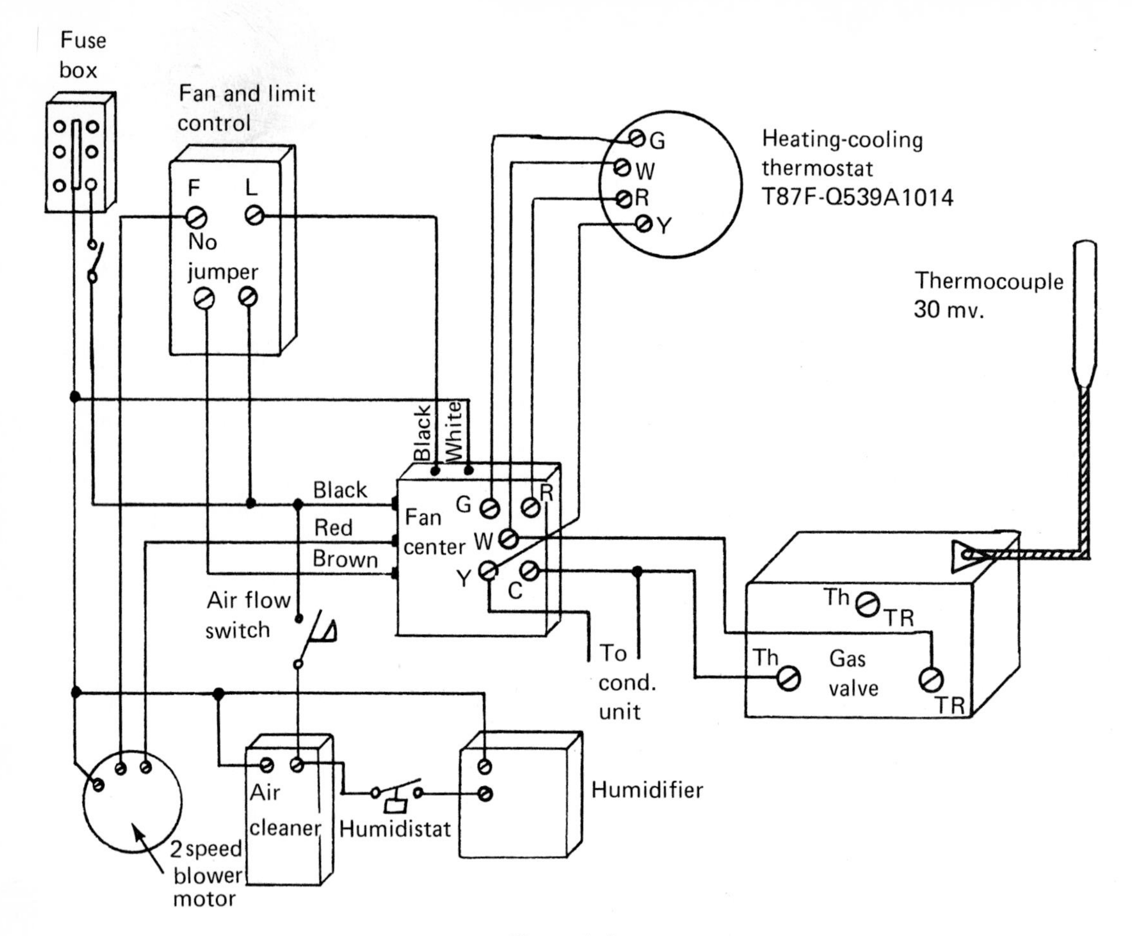
Fuse
box
Fan and limit
control
F
L
No
jumper
G
W
R
Y
Heating-cooling
thermostat
T87F-Q539A1014
Thermocouple
30 mv.
Black
White
Black
Red
Brown
Fan
center
G
R
W
Y
C
Air flow
switch
To
cond.
unit
Th
TR
Th
Gas
valve
TR
Air
cleaner
Humidistat
Humidifier
2 speed
blower
motor

Figure 8.4a.

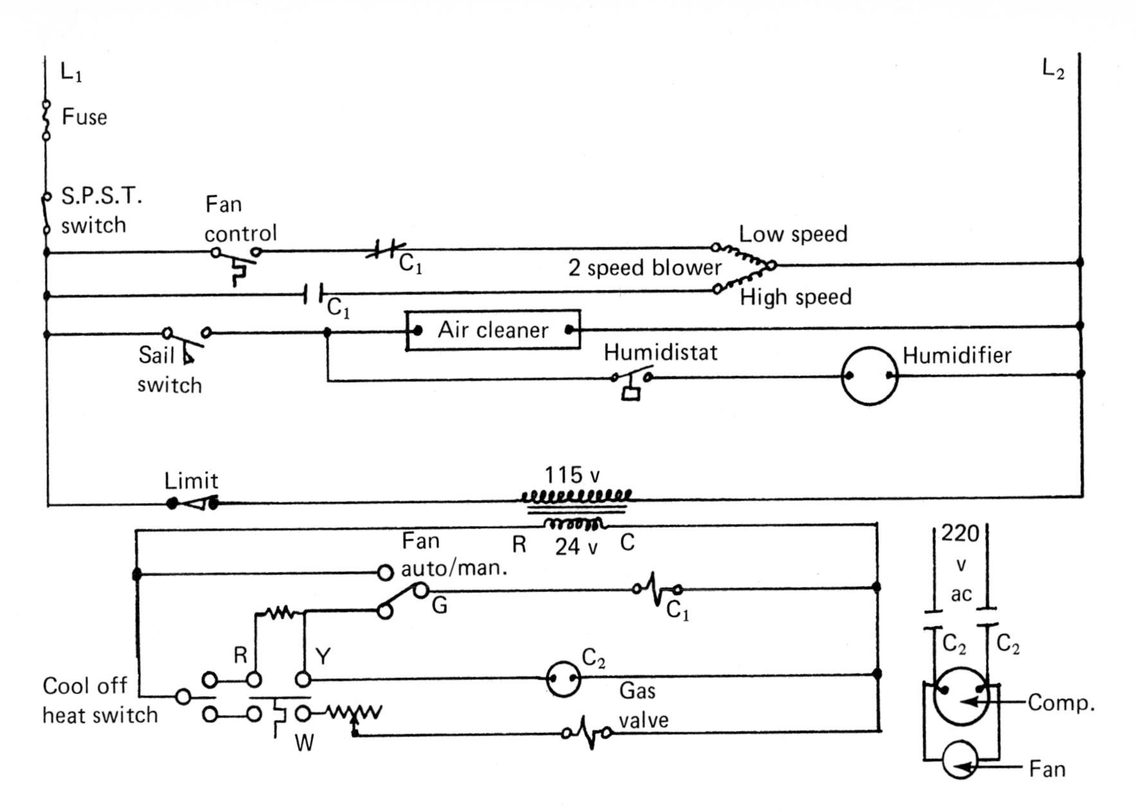
L1
L2
Fuse
S.P.S.T.
switch
Fan
control
C1
Low speed
2 speed blower
High speed
C1
Air cleaner
Sail
switch
Humidistat
Humidifier
Limit
115 v
Fan
auto/man.
R
24 v
C
220
v
ac
G
C1
R
Y
C2
C2
C2
Cool off
heat switch
Gas
valve
W
Comp.
Fan

Figure 8.4b.

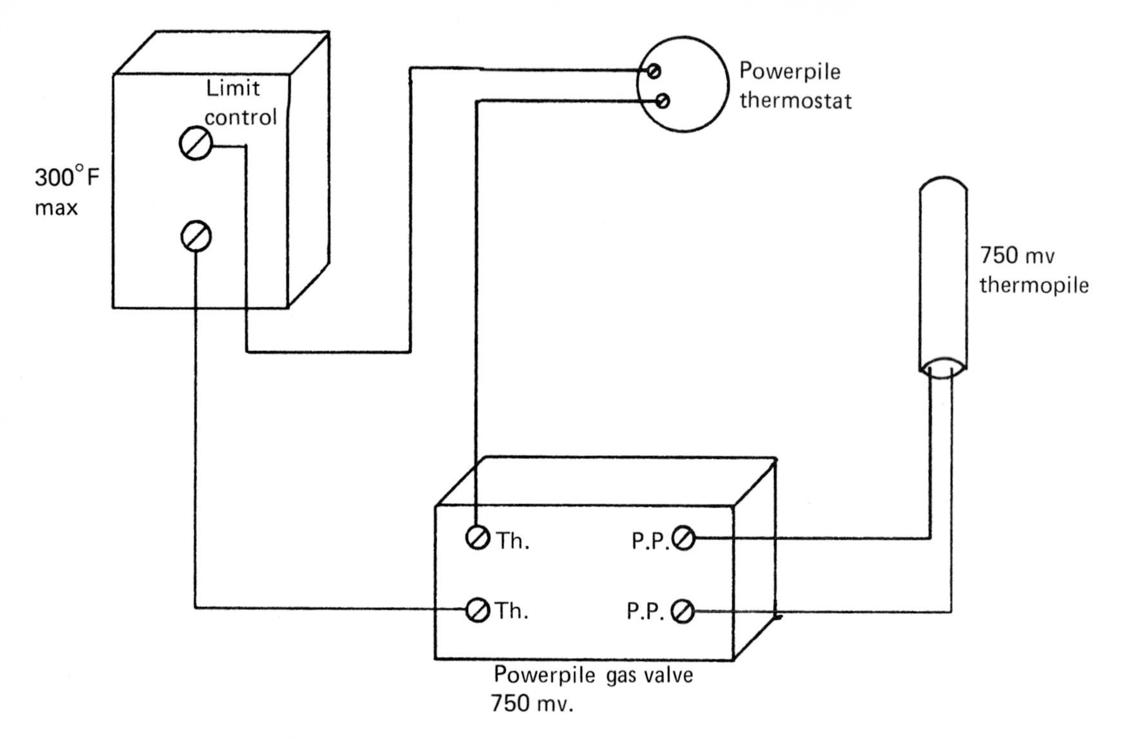

Figure 8.5a.

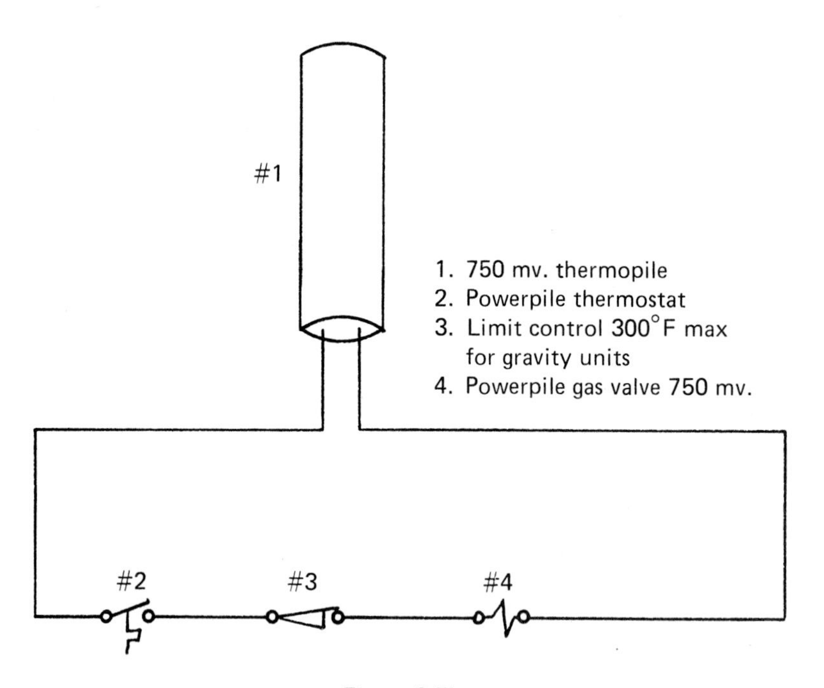

Figure 8.5b.

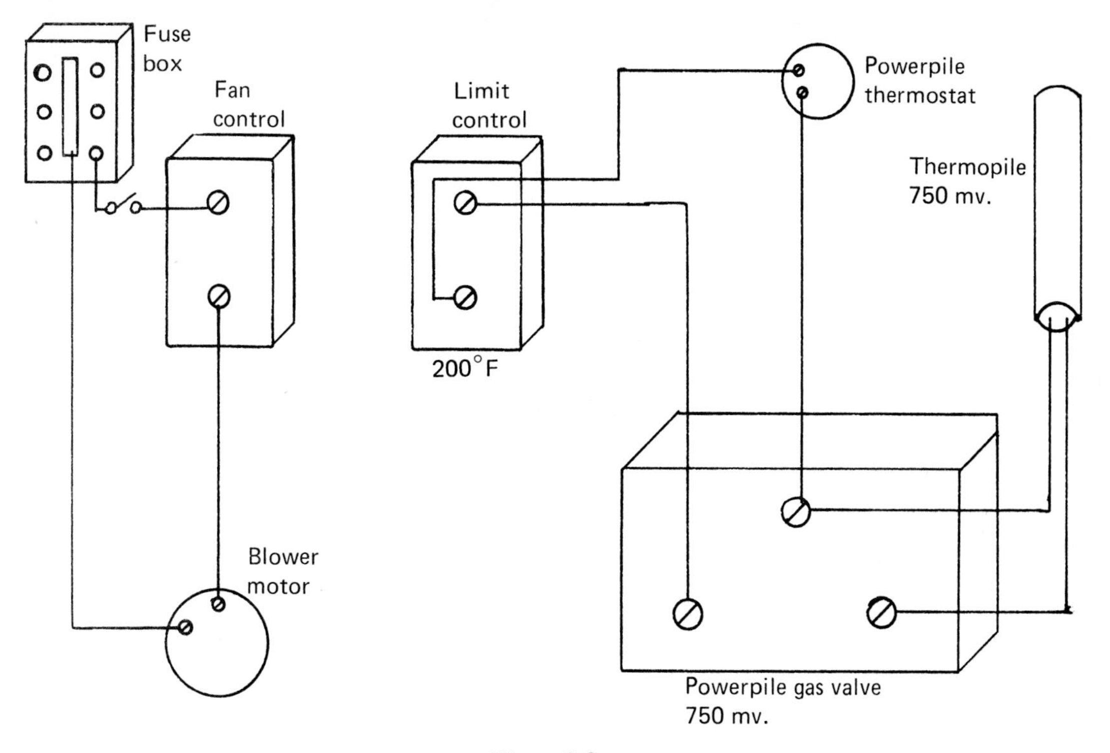

Figure 8.6a.

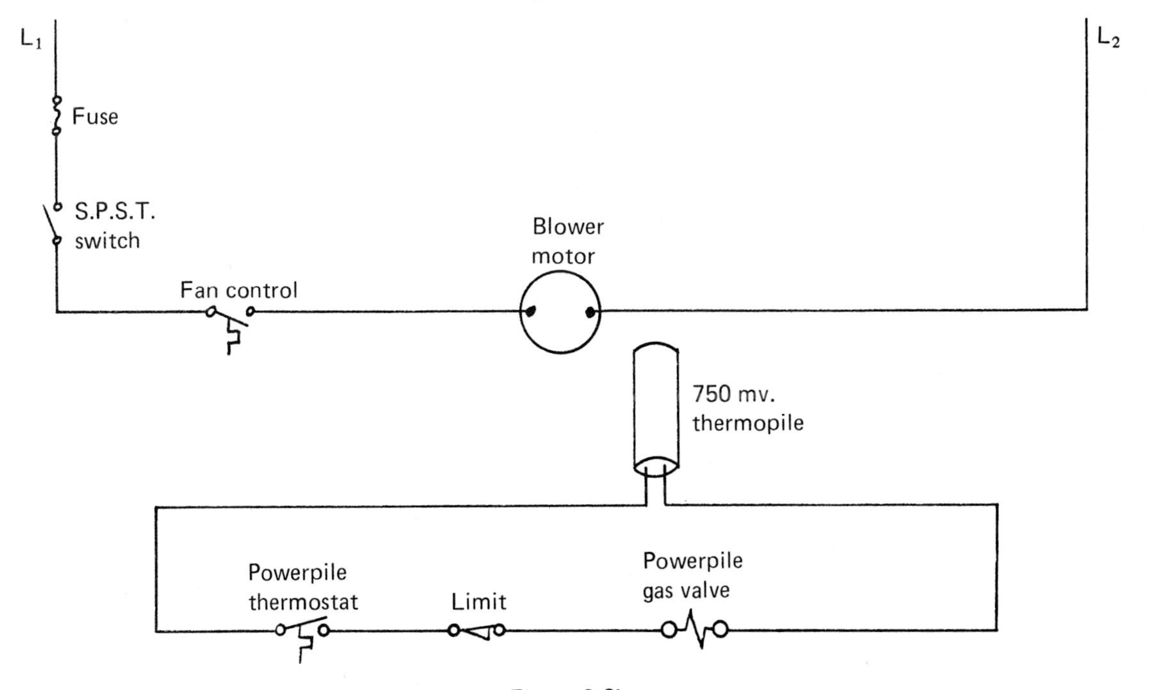

Figure 8.6b.

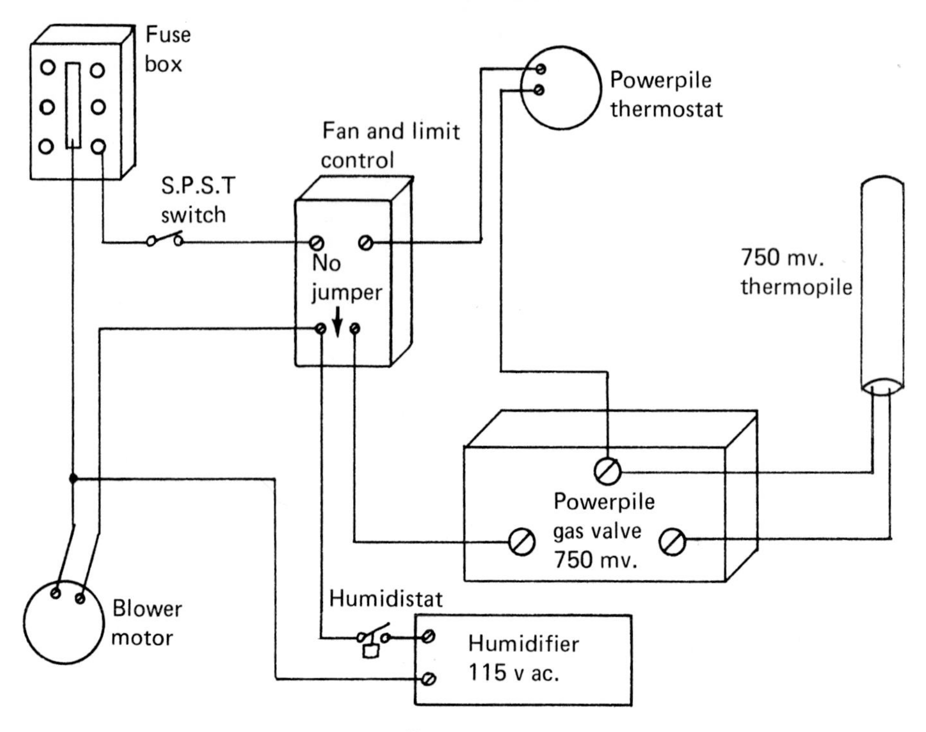

Figure 8.7a.

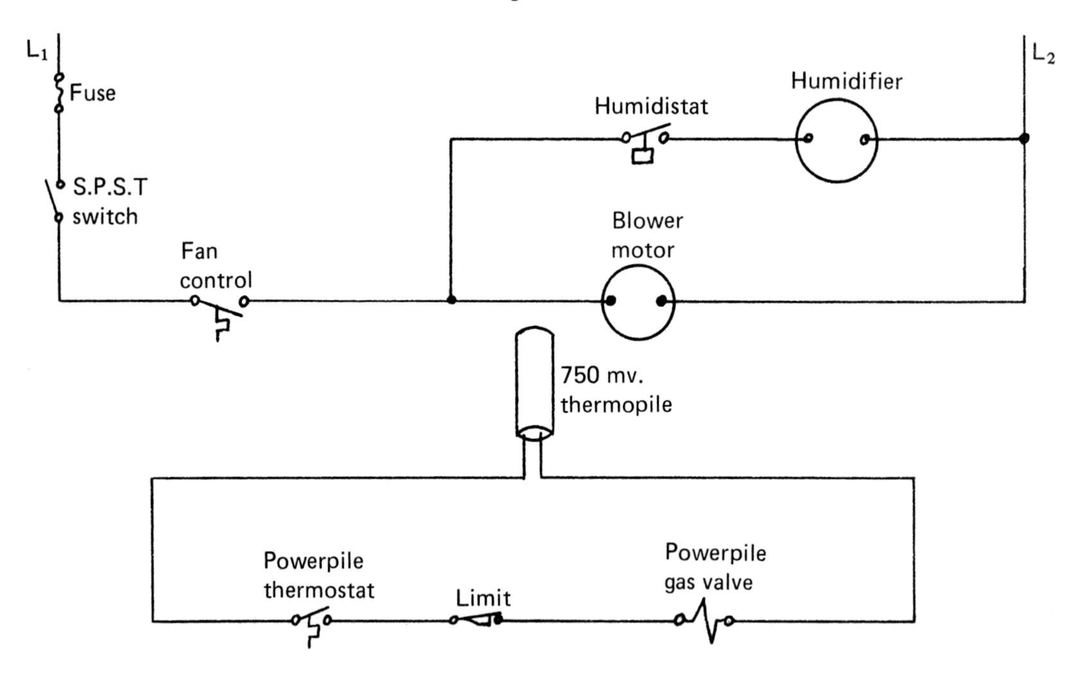

Figure 8.7b.

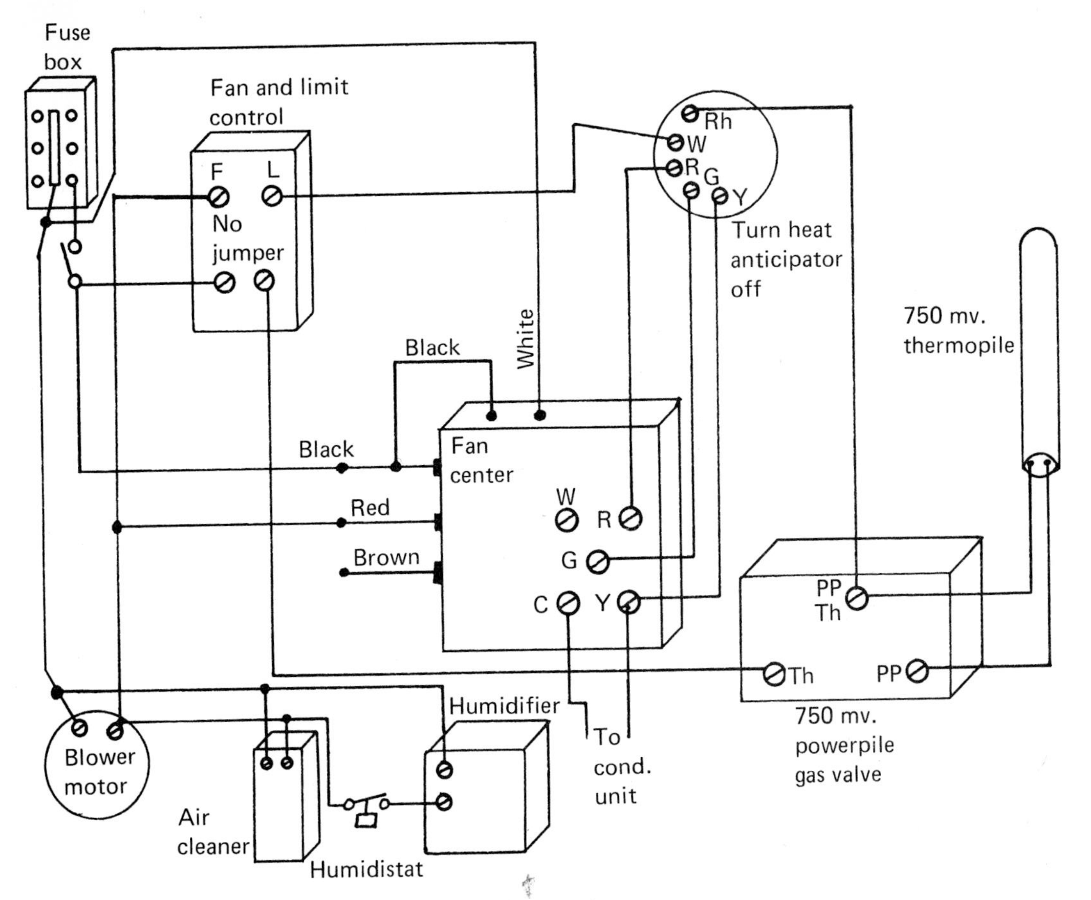
Fuse box
Fan and limit control
F
L
No jumper
Rh
W
R
G
Y
Turn heat anticipator off
750 mv. thermopile
Black
White
Black
Fan center
W
R
Red
G
Brown
C
Y
PP
Th
Th
PP
Humidifier
To cond. unit
750 mv. powerpile gas valve
Blower motor
Air cleaner
Humidistat

Figure 8.8a.

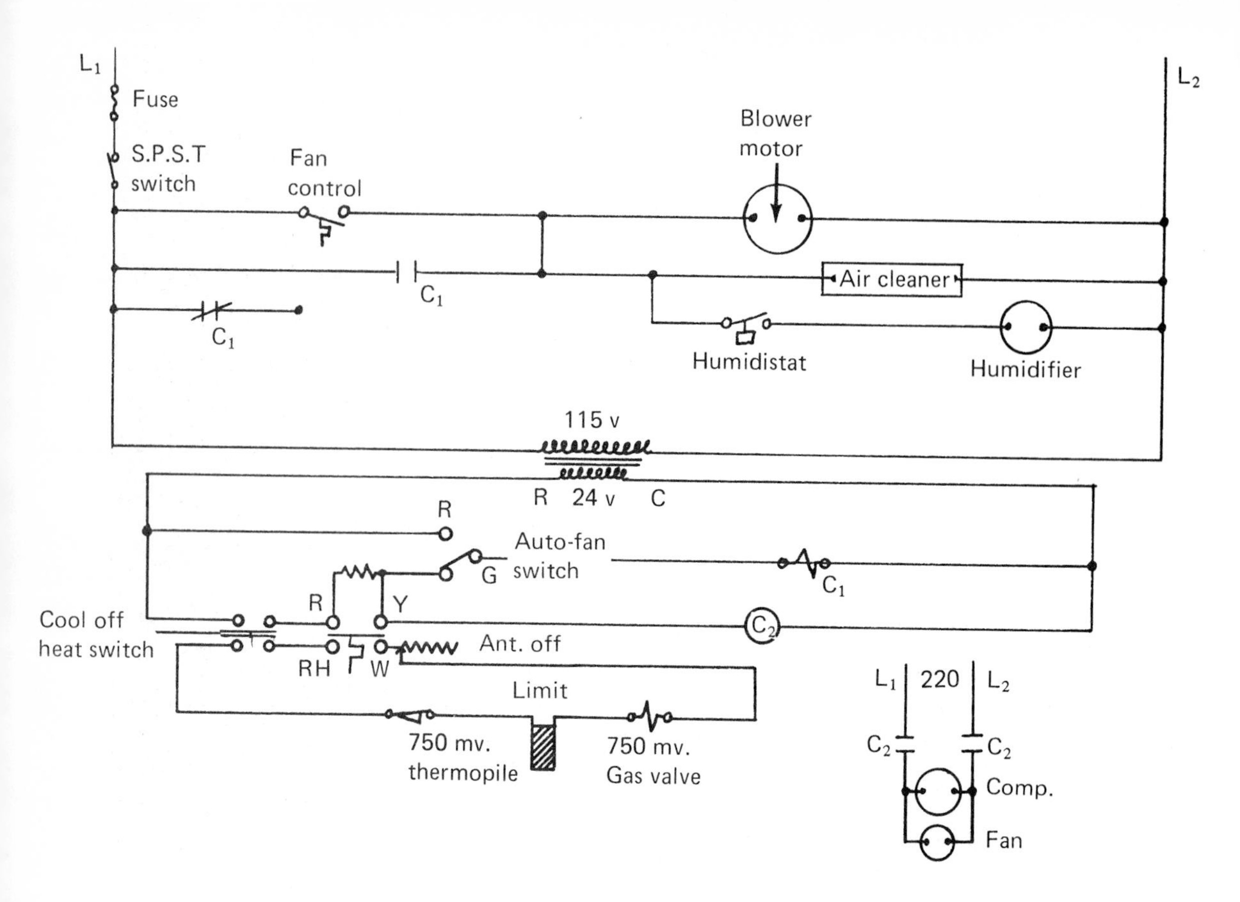
L1
L2
Fuse
S.P.S.T
switch
Fan
control
Blower
motor
C1
Air cleaner
C1
Humidistat
Humidifier
115 v
R 24 v C
R
Auto-fan
G switch
C1
R
Y
Cool off
heat switch
C2
RH
W
Ant. off
Limit
750 mv.
thermopile
750 mv.
Gas valve
L1
220
L2
C2
C2
Comp.
Fan

Figure 8.8b.

Chapter 9
ELECTRONIC AIR CLEANER

Electro-Air Slim Line

This dry type electronic air cleaner is designed for installation with any forced air, Residential furnace. Up to 95% of the dirty particles, carried to the electronic air cleaner by the duct system are removed by the unit. The slim line employs an Agglomerator principle which allows the collecting cell to be cleaned only once every 12 to 18 months. Dirt particles trapped by the collecting cell gather on a lightweight agglomerator pad which is cleaned more frequently. The unit pictured has an optional charcoal frame which provides effective odor removal.

Figure 9.1. Electro-air slim line.

The servicing of slim line electronic air cleaners is a simple procedure. The checkout steps listed are given in a logical concise manner. If they are followed in the manner described, locating the problem area will be easily accomplished.

The majority of service problems occur in the high voltage or secondary portion of the circuitry. The circuit is designed so that the faulty component can be easily detected and corrective measures taken in a minimum of time.

The circuits used in all slim line units are a standard application, called the voltage doubler. Its name means exactly what it implies—the output voltage of the power supply is twice that of the actual transformer output. In slim line units the ionizer and plate section are set at different voltages. The ionizer is 7,700 volts and the plate section carries 3,900 Volts. The plate section of the collecting cell serves as one of the capacitors in the voltage doubler circuit. The following illustration shows the circuit diagram for the slim line units. (See Figs. 9.2a and 9.2b.)

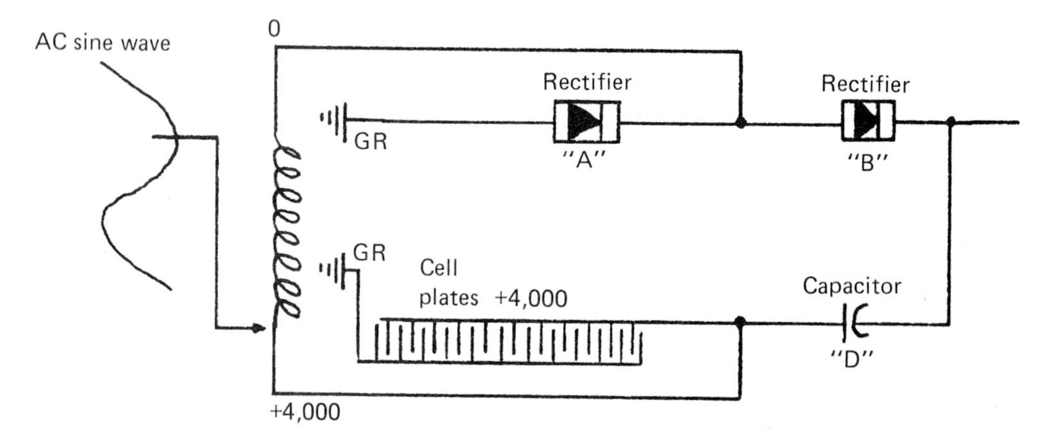

Figure 9.2a.

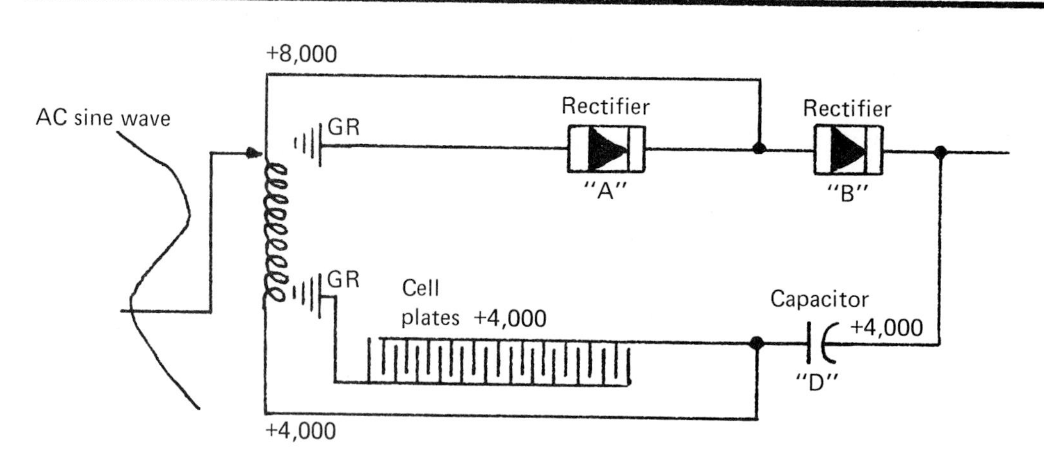

Figure 9.2b. Slim line voltage doubler circuit.

The two major component areas in the unit are the collecting cell and the power pack. Unless each of these component areas is correctly isolated and proven to be functioning properly, it is impossible to proceed with servicing in a logical manner. Before attempting to service the electronic air cleaner, the following steps should be taken to completely de-energize the unit and gain preliminary access to the interior.

1. Turn the **On/Off** switch to the **Off** position.
2. Turn the chrome handle a quarter turn to release the power pack from the air cleaner cabinet.
3. Pull the pack away from the air cleaner. This will accomplish three stages required for servicing.
   a. The 120 volt input to the power pack is disconnected.
   b. The high voltage to the collecting cell is broken.
   c. The power pack will be separated from the cabinet and collecting cell.

The unit is now completely de-energized. In order to service a slim line unit it is necessary to energize the various components of the units in the following manner to establish a service ready stage.

1. Place the power pack's **On/Off** switch in the **Off** position.
2. Using a standard household extension cord, plug the male portion into the nearest standard wall receptacle. If the furnace system has a continuous fan operation, the male plug can be inserted into the female receptacle at the top of the cell cabinet.
3. The other end of the extension cord is to be plugged into the receptacle at the back of the power pack.

The above two conditions will be repeatedly referred to in descriptions on how to service various component areas. The two basic steps should be repeated each time a service ready stage is indicated. (See Fig. 9.3)

The A.C. ammeter that is mounted on the front of the slim line power pack is the key component in determining if the electronic air cleaner is functioning correctly. The meter registers primary input current to the power pack and will show any of the following three conditions,—**Off—On—Service.** It is possible to completely check a slim line unit for service problems utilizing various readings that are recorded on the A.C. ammeter. However, professional service personnel will want to use either a kilovoltmeter-probe and/or a volt-ohmmeter.

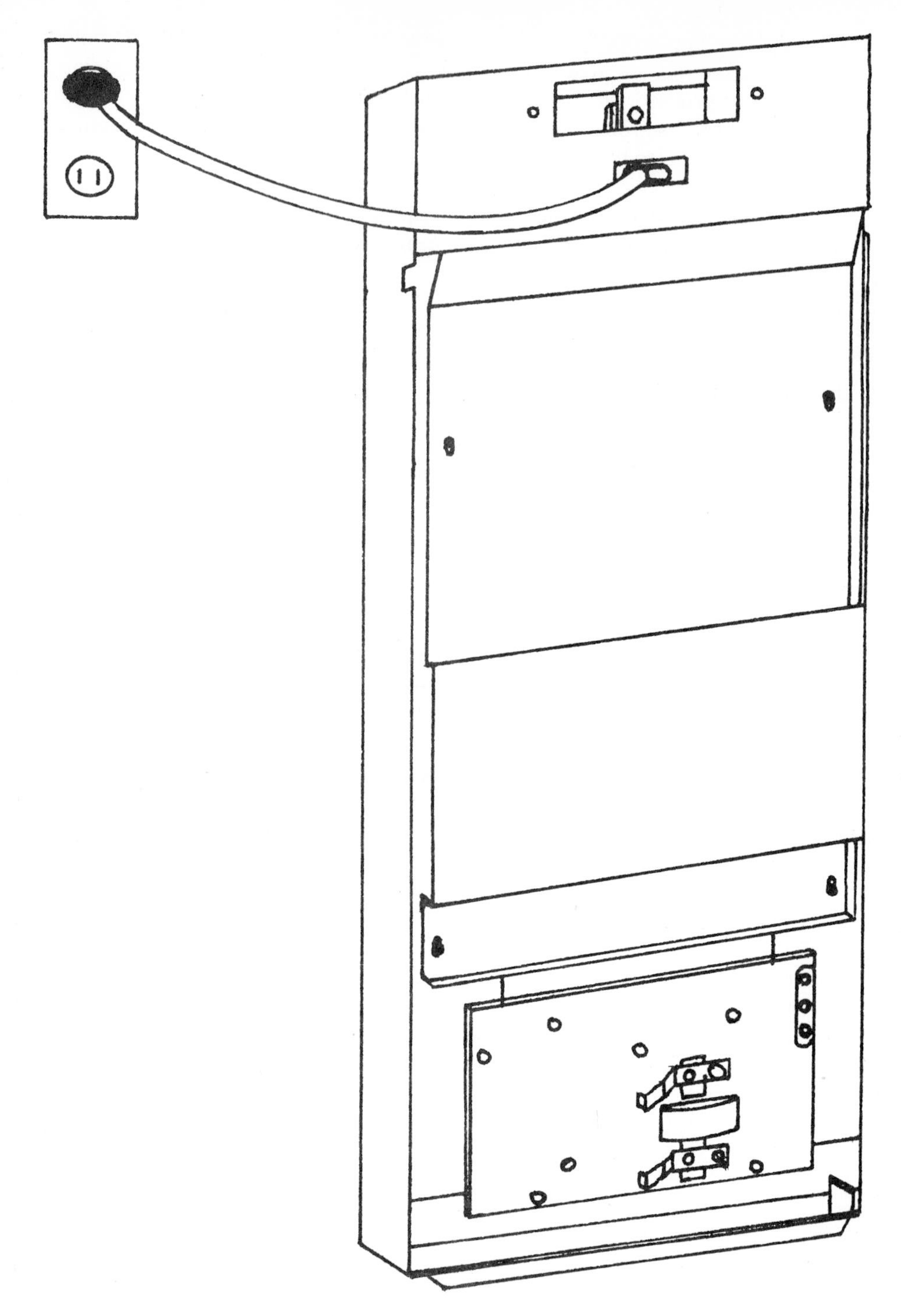

Figure 9.3. Ready service stage power pack.

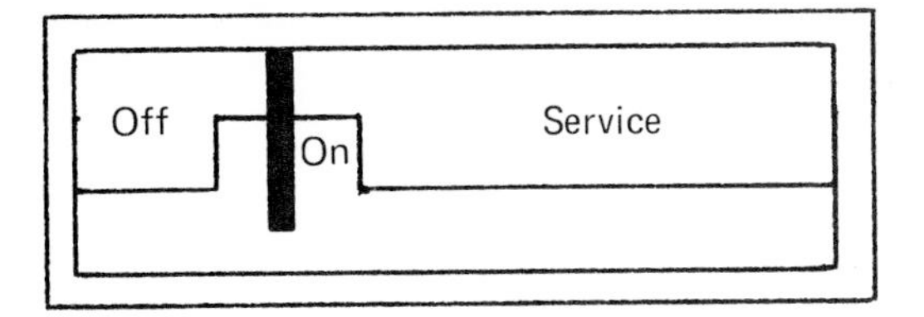

Figure 9.4.

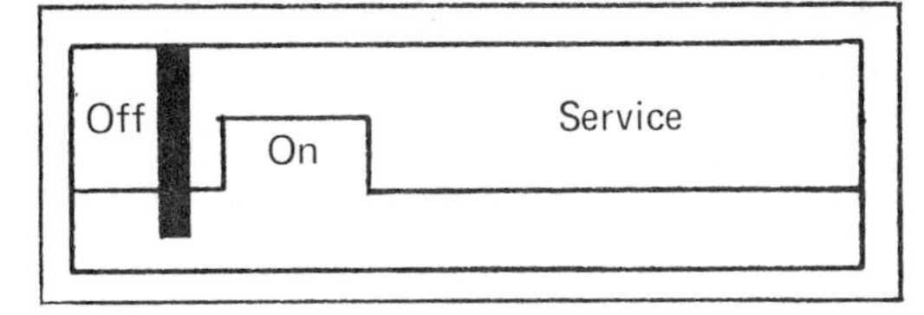

Figure 9.5.

The first visual confirmation that a unit is not functioning properly shown by the A.C. ammeter mounted on the front of the power pack. The indicating needle on the A.C. ammeter will be in the *black service area* to the right of the *white on area.* Because there are only two major component areas in the electronic air cleaner, the service problem has to be located in either the collecting cell or the power pack. Occasionally problems will occur in both areas at the same time. It is necessary to isolate the power pack and cell from each other and to test each component area independently. The first major component to be checked is the power pack.

1. Establish the service ready stage. (Extension cord providing power to power pack.)
2. Connect one lead from the substitute capacitor (See Fig. 9.3) to the bottom spring contact (See Fig. 9.6) of the power pack.
3. Connect the other lead from the substitute capacitor to the power pack housing.
4. Connect the kilovoltmeter-probe as shown in (Fig. 9.6). Turn the **On/Off** switch that is found on the front of the power pack to the ON position. Read the voltages recorded on the kilovoltmeter. If the kilovoltmeter voltages are below the 9,000 and 4,500 figures and the A.C. ammeter's indicating needle is in the **Service** area, there is a faulty high voltage component. If the readings are approximately 9,000 and 4,500, the problem is located in the cell. There are three major voltage components in the power pack. The easiest way to determine if each of these components is functioning correctly is to isolate each one and to check it separately. Access to the three major components inside the power pack is gained by removing the four sheet metal screws that hold the power pack back in place.

A. **Transformer Check**
Disconnect the red and yellow transformer leads from the red high voltage board. Turn the **On/Off** switch to the **On** position. Observe the A.C. ammeter. If the indicating needle returns to the *white* **On** *area,* the transformer is functioning correctly. If the reading remains in the service area to the right of the **On** area, the transformer is defective and must be replaced.

B. **Rectifier Test**
Access to the rectifiers is gained by removing the four sheet metal screws that hold the red high voltage board in place. Remove the rectifiers from their clip holders. Insert a pencil in the selenium end of the rectifiers (arrow points toward this end.) (See Fig. 9.8.) If the rectifier is good, the selenium will slide easily within the tube against the spring in the opposite end. If the rectifier is defective, the selenium will expand against the side of the tube and cannot be moved.

C. If the checking procedures listed before have still not located the source of trouble, replace the capacitor assembly.

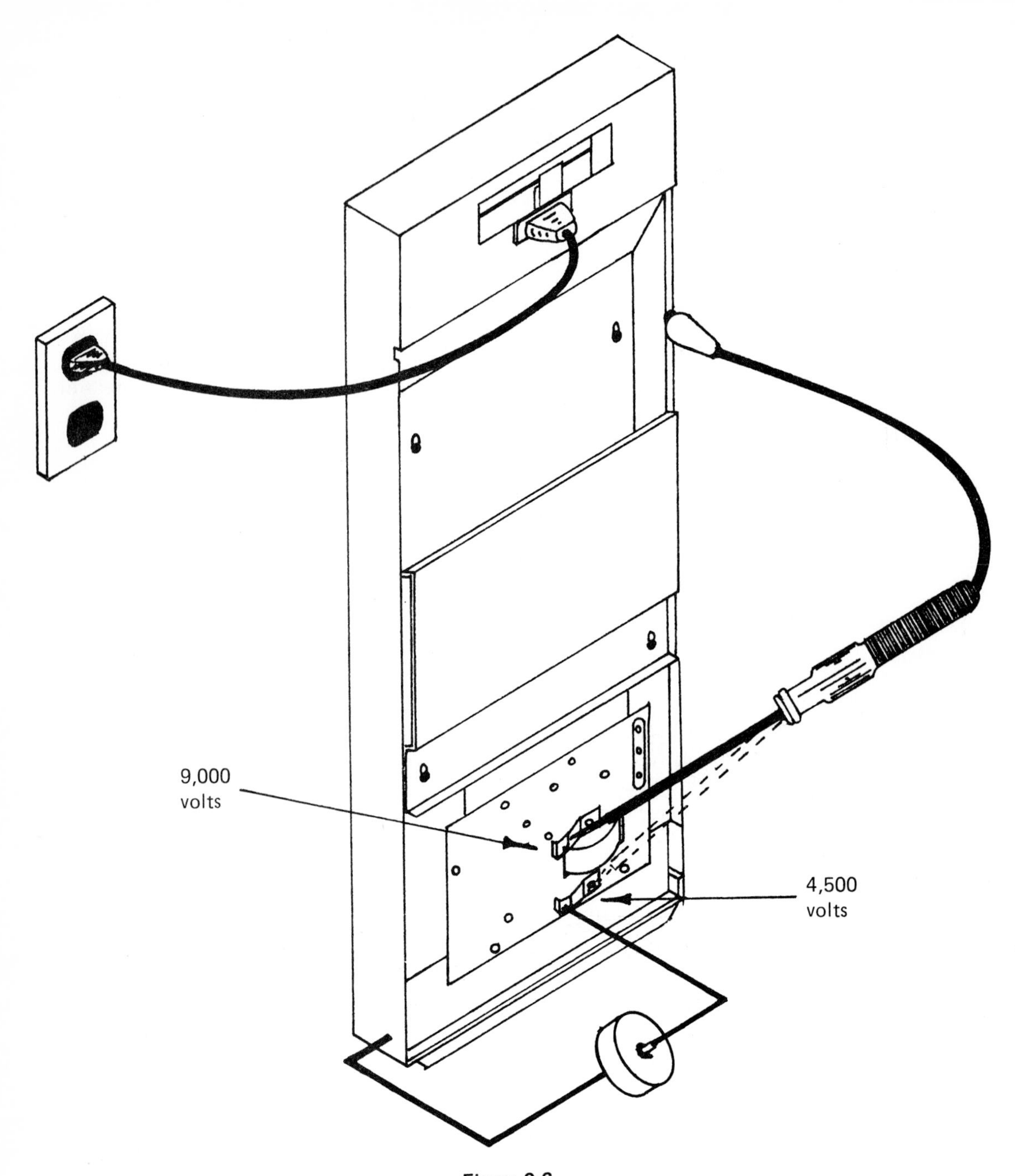

Figure 9.6.

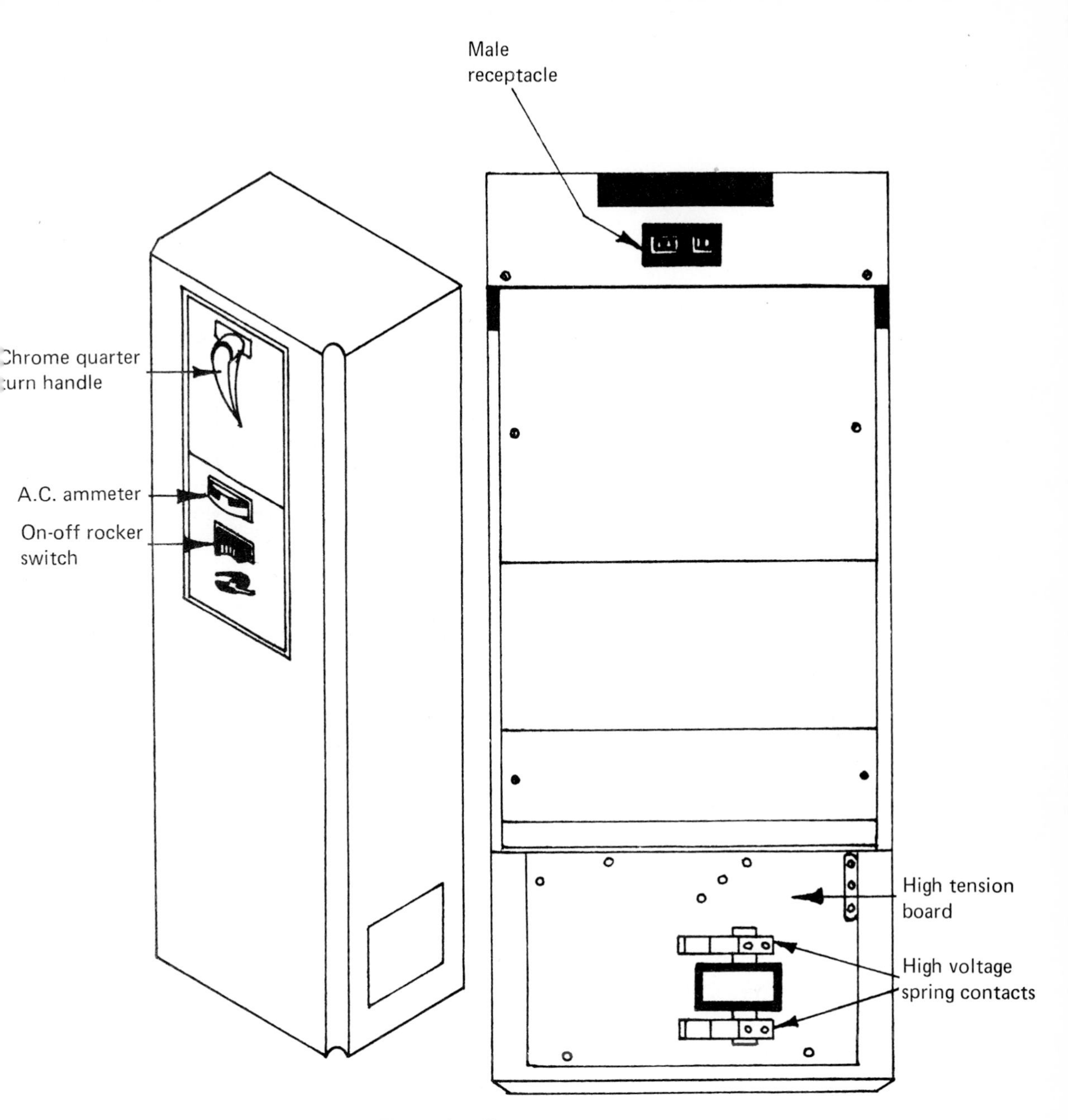

Figure 9.7. Slim line power pack.

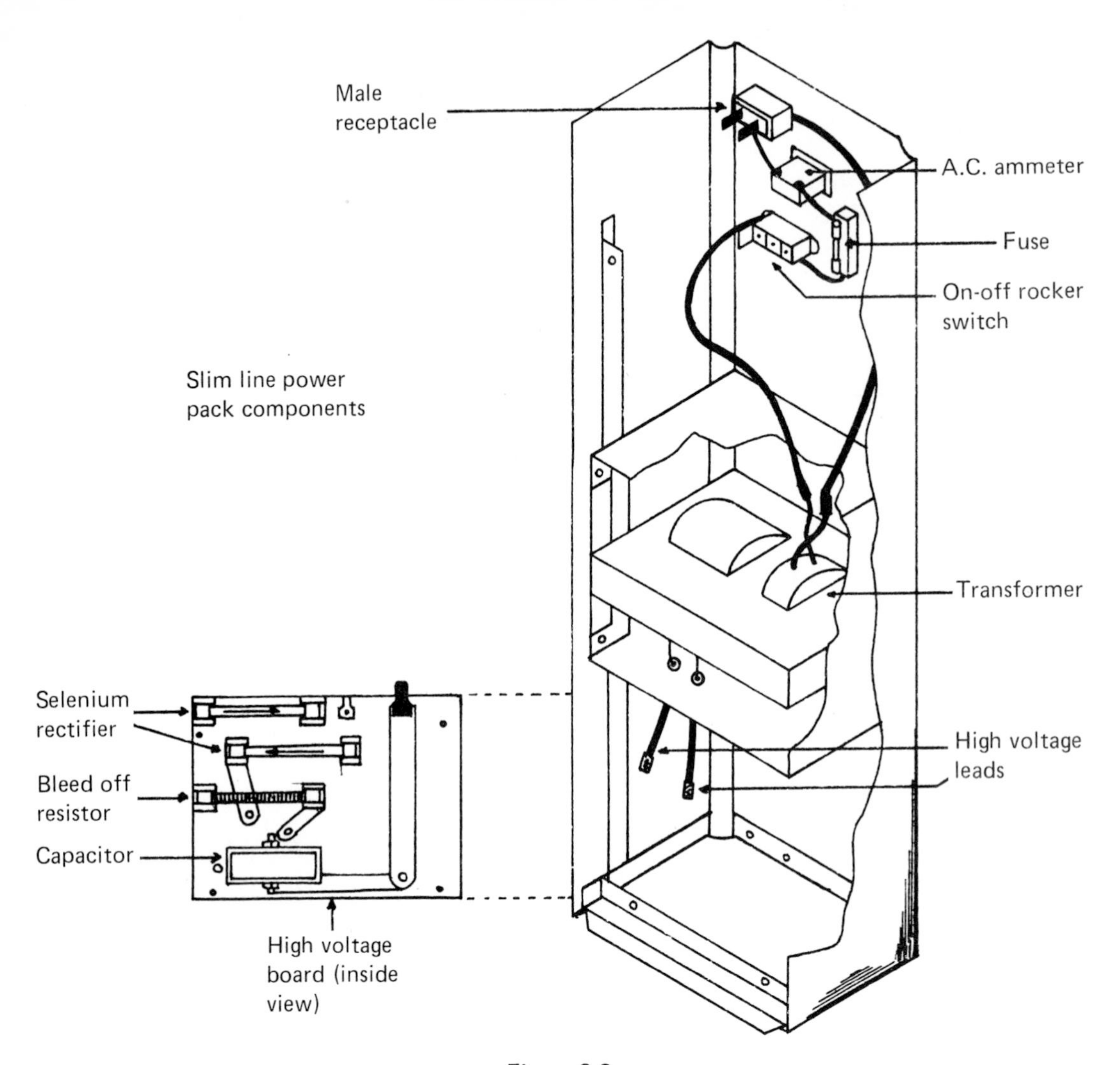

Figure 9.8.

Servicing the Collecting Cell

A. **Cell Plates**

1. Pull the collecting cell out of the electronic air cleaner and visually inspect for foreign particles that may have shorted the plates. Also inspect for bent collecting plates. If nothing is found wrong visually, a further check can be made with a volt-ohmmeter.
2. With the ohmmeter scale set at Rx 100,000, check two adajcent plates. A reading of infinite ohms should be obtained. Any reading but infinite ohms proves that some foreign object is lodged between the plates or the adjacent plates are touching one another.
3. If the procedures listed above have failed to locate the problem, thoroughly clean the cell. After the cell has dried, repeat the cell checkout procedures.

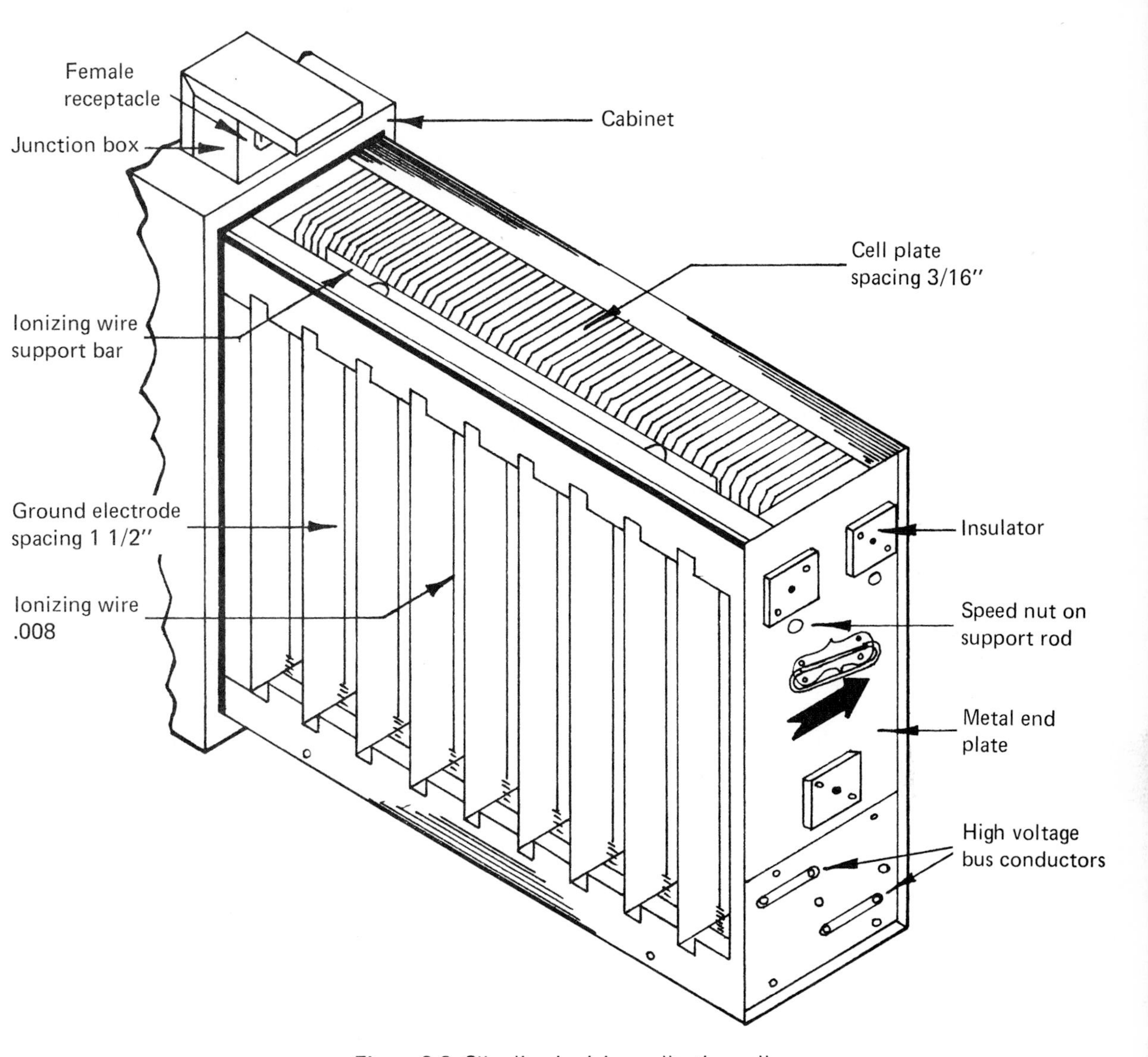

Figure 9.9. Slim line ionizing collecting cell.

B. **Ionizing Section**

If the service problem has not been located in the cell plates, the next area to investigate is the collecting cell's ionizing section.

1. Visually check and see if any of the porcelain insulators are cracked or broken. Also inspect for broken ionizing wires. If wire is broken, remove pieces and replace with a new wire.
2. In addition to visually checking the insulators, a kilovoltmeter-probe can also be employed to determine if any insulator is defective.
3. Establish the service ready stage. (Extension cord providing power to the power pack.)

4. Using three of the 6 ′ test leads, connect the cell to the power supply (See Fig. 9.10.)
5. Using a plastic handled screwdriver, remove the outside mounting screw that goes through the front frame into the insulator.
6. Move the insulator away from the front frame by pulling on the metal bar supporting the ionizing wire. This procedure will break any possible connection between the defective insulator and the cell.
7. Energize the power **On-Off** supply by turning the switch that is found on the front of the power pack to the ON postion. Read the voltage shown on the kilovoltmeter-probe.
   a. If the voltage recorded on the kilovoltmeter-probe reads in the normal (7,500-8,400) the insulator with the screw removed is defective and should be replaced with another insulator.
   b. If the voltage recorded on the kilovoltmeter-probe remains below the normal level (7,500-8,400) replace the screw and insulator that is being checked and proceed to another insulator where the same testing procedure should be applied. When the screw is removed from a defective insulator, the reading on the kilovoltmeter-probe will return to normal. This insulator should be replaced.

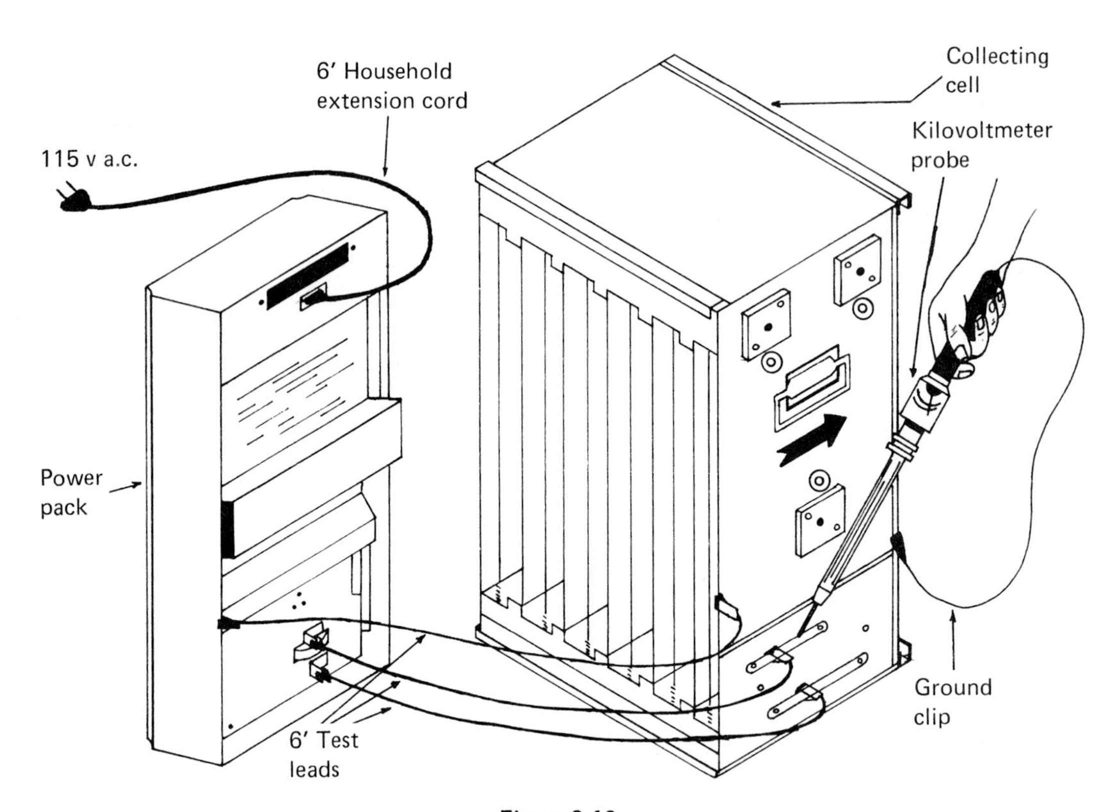

Figure 9.10.

8. Using the ohmmeter, a check for continuity can be made between the ionizing bar and the ground electrodes of the ionizing section. The reading should be infinite ohms on the Rx 100,000 scale. Reading but infinite ohms proves that a foreign object is in the ionizing section or that an ionizing wire has broken and is touching a ground electrode or there is a defective insulator. (The ohmmeter test is not totally conclusive. Occasionally, a very fine crack in an insulator will result in the insulator breaking down under a high voltage condition. In this particular case, the ohmmeter test for shorting would be inadequate.)

   Occasionally the power pack ammeter will indicate that the unit is operating correctly, when in fact, it is not. In this situation the indicating needle is registering in any portion of the *white on area.* To determine if the unit is functioning with the correct output voltage, it is necessary to check the voltages using the kilovoltmeter-probe. This procedure is as follows:

   a. Black clip from the kilovoltmeter-probe's lead is attached to the power pack housing. The metal end of the probe should touch the top spring contact. Readings on the kilovoltmeter-probe should be between 7,200 and 8,000 volts.
   b. If the voltages are lower than 7,200, proceed as follows:
      (1) Check the rectifiers to see that they match the directional arrows on the high voltage board.
      (2) Capacitor open-replace the capacitor.

   The first indication that a slim line unit is not working correctly is when the A.C. ammeter is in the *black off area.* This indicates that 120 volt power is not being supplied to the transformer. To determine that voltage is or is not available, proceed in the following manner:

   a. Check the supply voltage with a test light or with a volt-ohmmeter. Set the ohmmeter on the 0-300 v scale. It should read 110-120 A.C. If the light continues to glow, indicating an adequate power supply, proceed to individually check the primary components.
   b. Check each of the primary components shown with a test light. The test light will glow until you pass the defective component. An alternate method of checking the primary components in the power pack to discover a defective component is to use a volt ohm-meter. Before attempting this method, remove the extension cord from the power pack. The components to be checked with a volt-ohmmeter are as follows:
      (1) Fuse—Remove the fuse and check for continuity with the volt-ohmmeter on the ohms scale. The reading should be zero ohms on the Rx1 scale. If the reading is zero, the fuse is good and should be removed for the rest of the tests. If the fuse is bad, replace it.
      (2) **On-Off Switch**—Connect the volt-ohmmeter across the switch and operate the switch several times. The meter should deflect to zero ohms on the Rx1 scale when the switch is on and to infinite ohms on the Rx 100,000 scale when it is off.
      (3) **A.C. Ammeter Open**—With the volt-ohmmeter on the Rx1 scale, connect the leads of the meter across the terminals of the A.C. ammeter. A reading of approximately 2.5 ohms indicates that the meter is operating properly. A

reading of infinite ohms shows that the meter's winding is open and the meter should be replaced.

(4) An alternate check of the ammeter can be performed by using the kilovoltmeter-probe. Replace the extension cord and the fuse on the power pack. Short out the meter (connect a wire from one terminal to the other) and turn the **On-Off** switch that is found on the front of the power pack to the **On** position to energize the unit.

Use the kilovoltmeter-probe to determine if voltage has been restored to the output of the power pack. If the voltage reads correctly (7,200 to 8,000), the A.C. ammeter is defective and should be replaced.

Determining that the collecting cell is functioning after correcting a power pack service problem. There is a possibility that two separate service problems may have occurred in the electronic air cleaner: one located in the power pack, and the other in the collecting cell. If the problem has been located in the power pack and has been satisfactorily corrected, the collecting cell should be checked to determine that there is no problem there. This is done in the following manner:

1. Connect the power supply to the collecting cell as shown in (Fig. 9.10)
2. Connect the kilovoltmeter-probe by attaching the black clip lead to the power pack housing and touch the metal end to the top spring contact.
3. Turn the **On-Off** switch found on the front of the power pack to the **On** position. Observe the A.C. ammeter and the kilovoltmeter-probe meter. If the indicating needle of the A.C. ammeter is in the *black service area* to the right of the *white on area,* and the kilovoltmeter-probe indicates less than 7,200 volts, the collecting cell contains a service problem.
4. Proceed to check out the collecting cell in the manner described in servicing the cell.

Determining if the A.C. ammeter is defective. The check-out procedure previously given for the A.C. ammeter is used to determine that the ammeter is in the open state and will not allow the primary outlet to flow. There is a slight possibility that the ammeter would indicate that the electronic air cleaner requires service, when in fact, the ammeter itself is defective. This could be caused by poor workmanship in the manufacturing of the ammeter or user tampering. The ammeter in this condition would show the indicating needle in the service area. To check out the ammeter, the following procedure should be used:

1. Establish the service ready stage. (Extension cord providing to power pack.)
2. Using the kilovoltmeter-probe check the voltage to the cell. Attach black clip lead from kilovoltmeter-probe to the power pack housing and touch metal end of probe to the top spring contact.
3. Turn the **On-Off** switch found on the front of the power pack to the **On** position to temporarily energize the air cleaner.
4. Observe the voltage shown on the kilovoltmeter-probe. If the voltage reading is normal (7,200 to 8,000 volts) the problem is a defective ammeter, which should be replaced with a new ammeter.

Additional Service and Maintenance Tips

1. **Initial Installation**
   Upon initial energizing of a unit a low voltage and high current condition may exist. This is due to the high ionization current draw of the new cell. This is a normal condition and will correct itself within approximately 48 hours of operation.
2. **Continuous Arcing**
   This is usually caused by bent or loose plates. It can be corrected by straightening plates and reconnecting plates to bars.
3. **Maintenance of Collecting Cells**
   Cells should be removed from the cabinet at the beginning of each heating season and manually washed with hot water and a dish washing detergent or DAX-W. After washing with the detergent, rinse with hot water and drain. Replace the cells and allow the furnace blower to operate approximately one hour before re-energizing the power pack. (Note: A regular garden hose is sufficient for this cleaning procedure.)

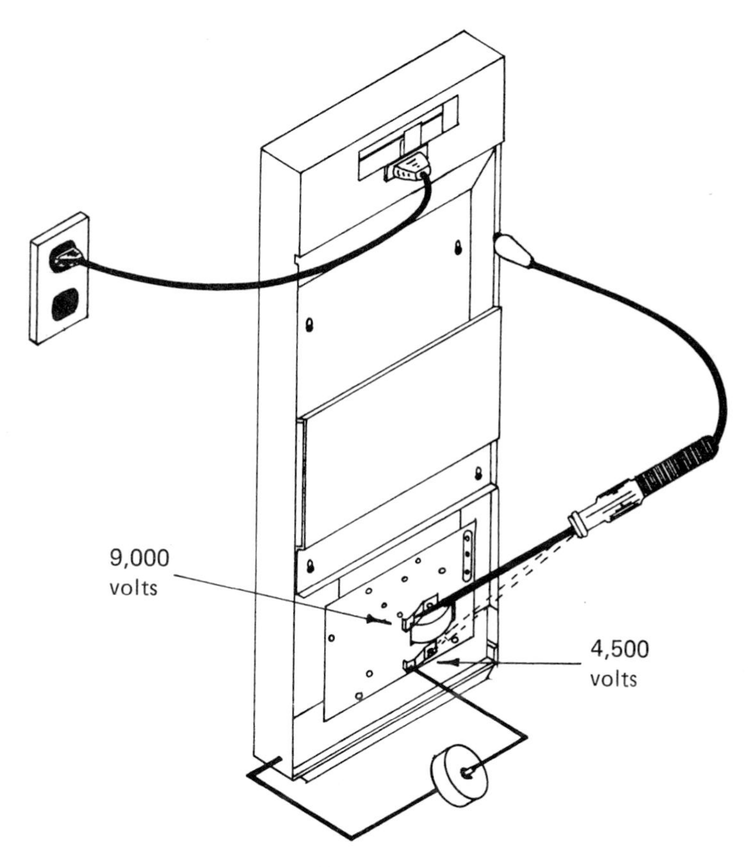

Figure 9.11.

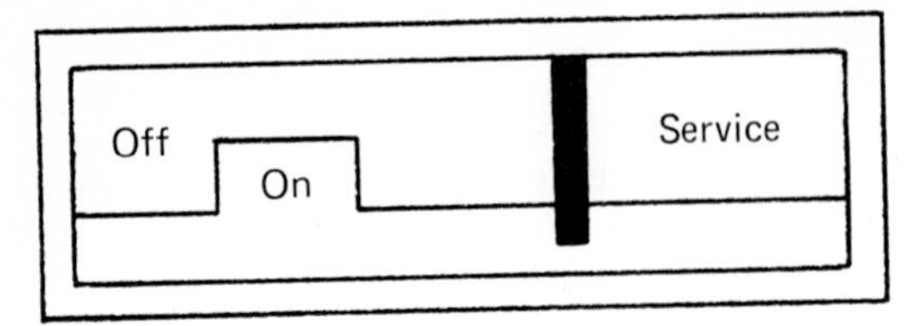

Figure 9.12.

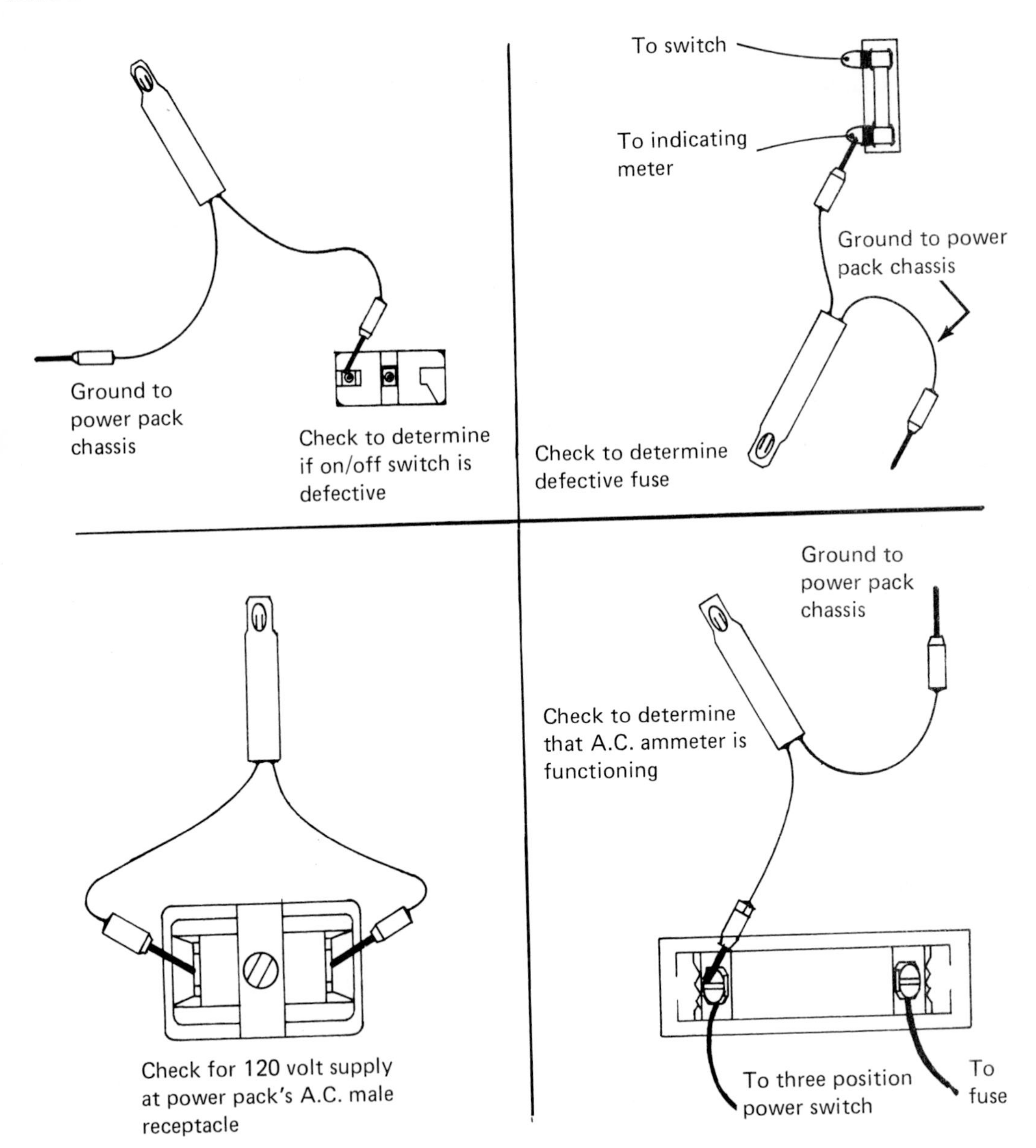

Figure 9.13.

Servicing Electro-Air Super Compact Units

The majority of the service problems occur in the high voltage or secondary portion of the circuitry. The circuit is designed so that the faulty component can be easily detected and corrective measures taken—in a minimum time. The circuits used in this residential unit are a standard application called the "voltage doubler." Its name means exactly what it implies—the output voltage of the power supply is twice that of the actual transformer output.

Figure 9.14. Super compact.

How The Voltage Doubler Circuit Works. On the first half of the A.C. voltage wave, the transformer secondary is polarized, (See Fig. 9.15). The rectifier "A" in effect connects its end of the transformer secondary to "Ground." The other end of the secondary is at +4,000 volts which voltage (electrical pressure) charges the "C" capacitor. On the second half of the A.C. wave, the voltages are reversed as shown below. The capacitor "C" still retains most of its 4,000 volt charge which raises the lower end of the secondary above ground by that amount. The direction of the polarity in the secondary now places the upper end of the winding at 4,000 volts above the lower end. This brings the upper end to 4,000 plus 4,000, or 8,000 volts above "ground." This is the voltage required for the ionizing field of the cell, and passes through rectifier "B" and capacitor "D." The two major component areas in the unit are the collecting cell and the power pack. Unless each of these component areas is correctly isolated and proven

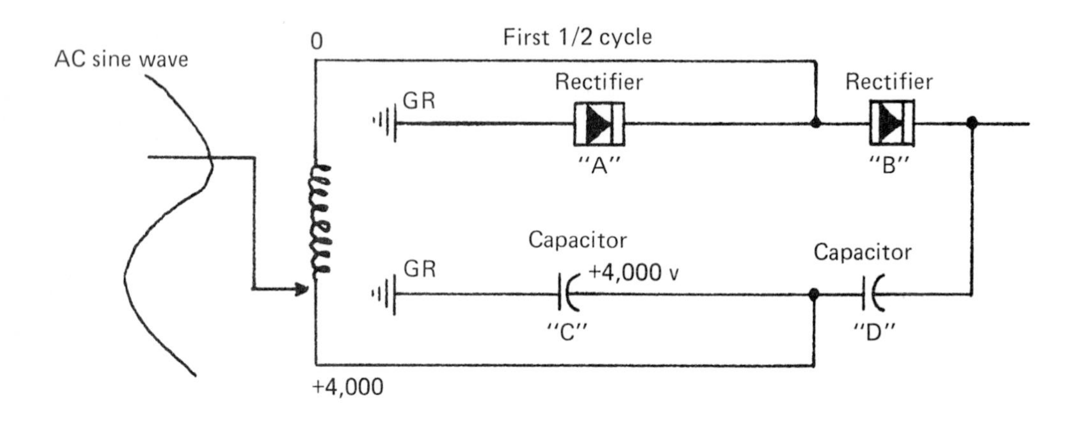

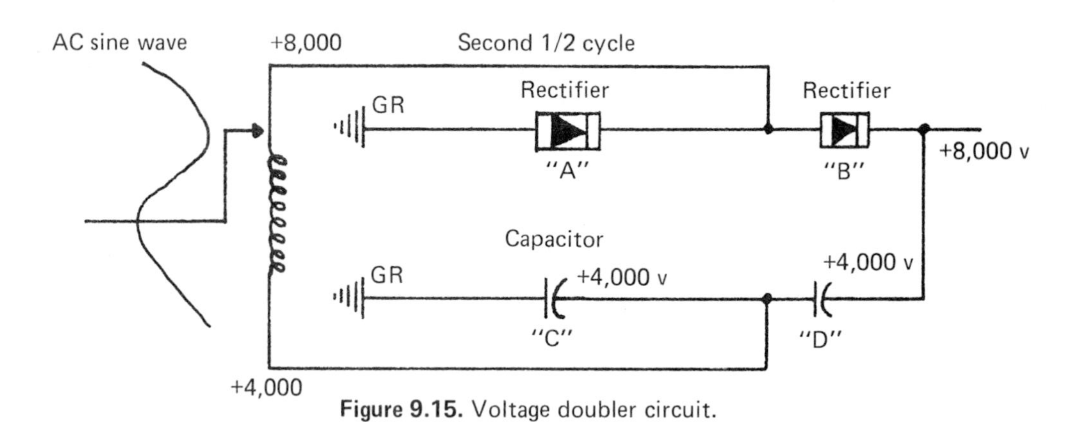

Figure 9.15. Voltage doubler circuit.

to be functioning properly, before attempting to service the electronic air cleaner, the following steps should be taken to completely de-energize the unit and gain access to the interior:

1. Turn the voltage selector switch to the **Off** position.
2. Remove the power supply cord from the top of the power pack.
3. Turn the quarter turn fastener to allow removal of the power pack lid.

Service Ready Stage

The unit is now completely de-energized with the interior parts exposed for servicing. In order to service the unit, it is necessary to energize the various components in the following manner to establish a service ready stage.

1. Reconnect the power supply cord to the male receptacle.
2. Turn the voltage selector switch to position #3 (see Figure 9.16).

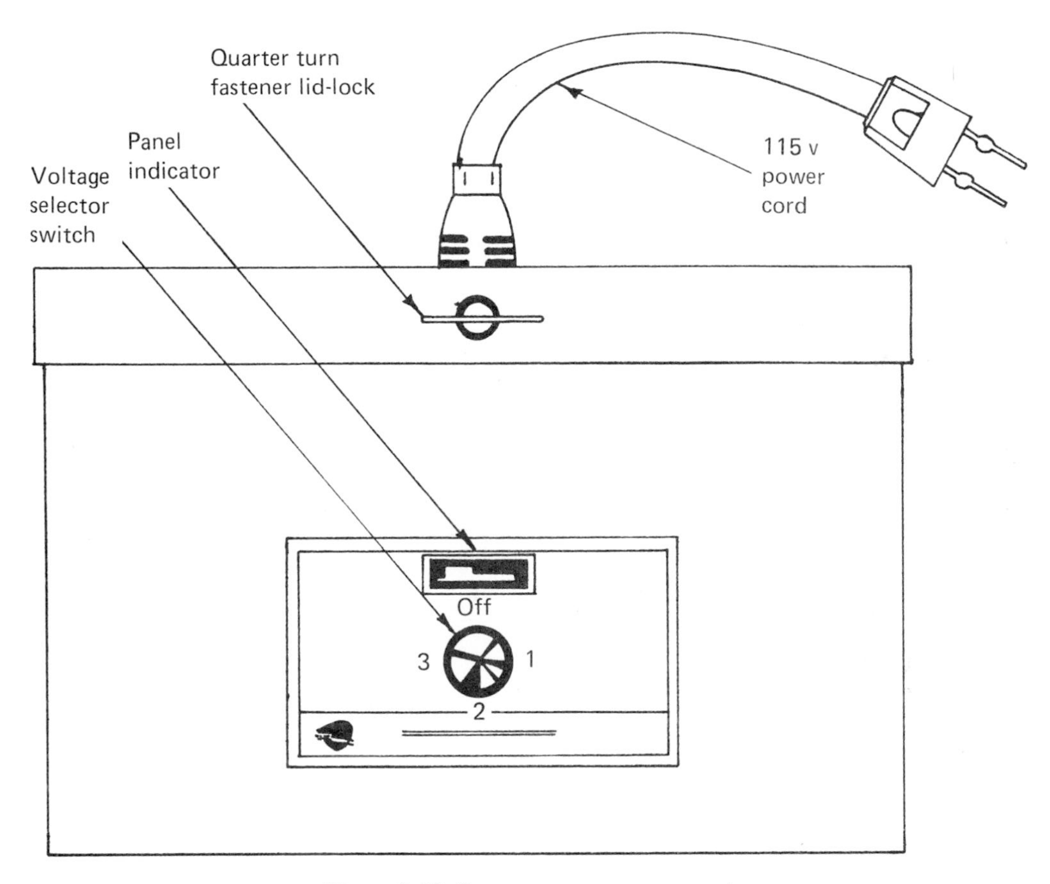

Figure 9.16. Super compact power pack.

3. To energize the power supply for a temporary period, depress the safety switch with thumb, exercising caution so that fingers do not make contact inside the power pack. This permits a by-pass of the safety switch that is part of the power pack. The above three conditions will be repeatedly referred to in descriptions on how to service various component areas. The three basic steps should be repeated each time a service ready stage is indicated. The A.C. ammeter that is mounted on the front of the super compact power pack is the key component in determining if the electronic air cleaner is functioning correctly. The meter registers primary input current to the power pack and will show any of the following three conditions: **On—Off—Service.**

The first visual confirmation that a unit is not functioning properly is shown by the A.C. ammeter mounted on the front of the power pack. The indicating needle on the A.C. ammeter will be in the *black service area* to the right of the *white on area.* Because there are only two major component areas in the electronic air cleaner, the service problem has to be located in either the collecting cell or the power pack. Occasionally problems will occur in both areas at the same time. It is necessary to isolate the power pack and cell from each other and to test each component area independetly. The first major component to be checked is the collecting cell.

1. Establish the service ready stage. Connect power supply cord, voltage selector switch turned to position #3 and safety switch ready to be depressed to temporarily energize the air cleaner.
2. Disconnect the high tension lead from the thru-bushing.
3. Attach the black clip from the kilovoltmeter-probe's lead to the power pack housing. The metal end of the probe should touch the metal bracket extending from the red high voltage board mounted inside the power pack. Caution should be exercised so that the unconnected high voltage lead does not come in contact with any metal portion of the power pack.
4. Depress the safety switch to temporarily energize the air cleaner.
5. Observe the voltage shown on the kilovoltmeter-probe. If it is 9,500 volts or higher, the problem is located in the collecting cell. If a kilovoltmeter-probe is not available, it is possible to determine the defective air cleaner component using the A.C. ammeter that is mounted on the front of the power pack.

If the ammeter's indicating needle returns to a position that is very low in the *white on area,* the problem is located in the collecting cell.

1. Disconnect the kilovoltmeter-probe and power supply cord from the power pack.
2. Loosen the two round-head machine screws mounted in the front of the cabinet, which support the power pack. This will permit the lifting of the power pack off the screws.
3. Unscrew and remove the thru-bushing from the cabinet door hole.
4. Totally remove the two cabinet door round-head machine screws that were previously loosened.
5. Remove cabinet door.

6. The collecting cell for the electronic air cleaner is held in place during shipment by a steel plate held in position by two screws and retaining nuts. If this shipping plate has not been removed, do so. It is not necesary to replace this plate once the unit is installed. The cell can now be easily slid out of the electronic air cleaner.

A. **Cell Plates**

1. Pull the collecting cell out of the electronic air cleaner and visually inspect for foreign particles that may have shorted the plates. Also inspect for bent collecting plates. If nothing is found wrong visually, a further check can be made with a volt-ohmmeter, not to be confused with a kilovoltmeter-probe.
2. With the ohmmeter scale set at Rx 100,000, check two adjacent plates. A reading of infinite ohms should be obtained. Any reading but infinite ohms proves that some foreign object is lodged between the plates or the adjacent plates are touching one another.
3. If the procedures listed above have failed to locate the problem, thoroughly clean the cell. After the cell has dried, repeat the cell check-out procedures.

B. **Ionizing Section**

If the service problem has not been located in the cell plates, the next area to investigate is the collecting cell's ionizing section.

1. Visually check and see if any of the porcelain insulators are cracked or broken. Also inspect for broken ionizing wires. If a wire is broken, remove pieces and replace with a new wire.
2. In addition to visually checking the insulators, a kilovoltmeter-probe can also be used to determine if an insulator is defective.
3. Establish the service ready stage. Connect power supply cord, voltage selector switch turned to position #3 and safety switch ready to be depressed to temporarily energize the air cleaner.
4. Using two of the 6 ′ test leads, connect the cell to the power supply as follows:
   a. one clip of the test should be connected to the power pack housing with the other end connected to the metal end of the collecting cell.
   b. The second test lead should be connected from the high voltage bracket to the mounting nut for the thru-bushing on the collecting cell.
5. Using a plastic handled screwdriver, remove the outside mounting screw that goes through the front frame into the insulator.
6. Move the insulator away from the front frame by pulling on the metal bar supporting the ionizing wire. This procedure will break any possible connection between the defective insulator and the cell.
7. Energize the power supply by depressing the safety switch. Read the voltage shown on the kilovoltmeter-probe.
   a. If the voltage recorded on the kilovoltmeter-probe reads in the normal (7,500-8,400) the insulator with the screw removed is defective and should be replaced with another insulator.
   b. If the voltage recorded on the kilovoltmeter-probe remains below the normal level (7,500-8,400).

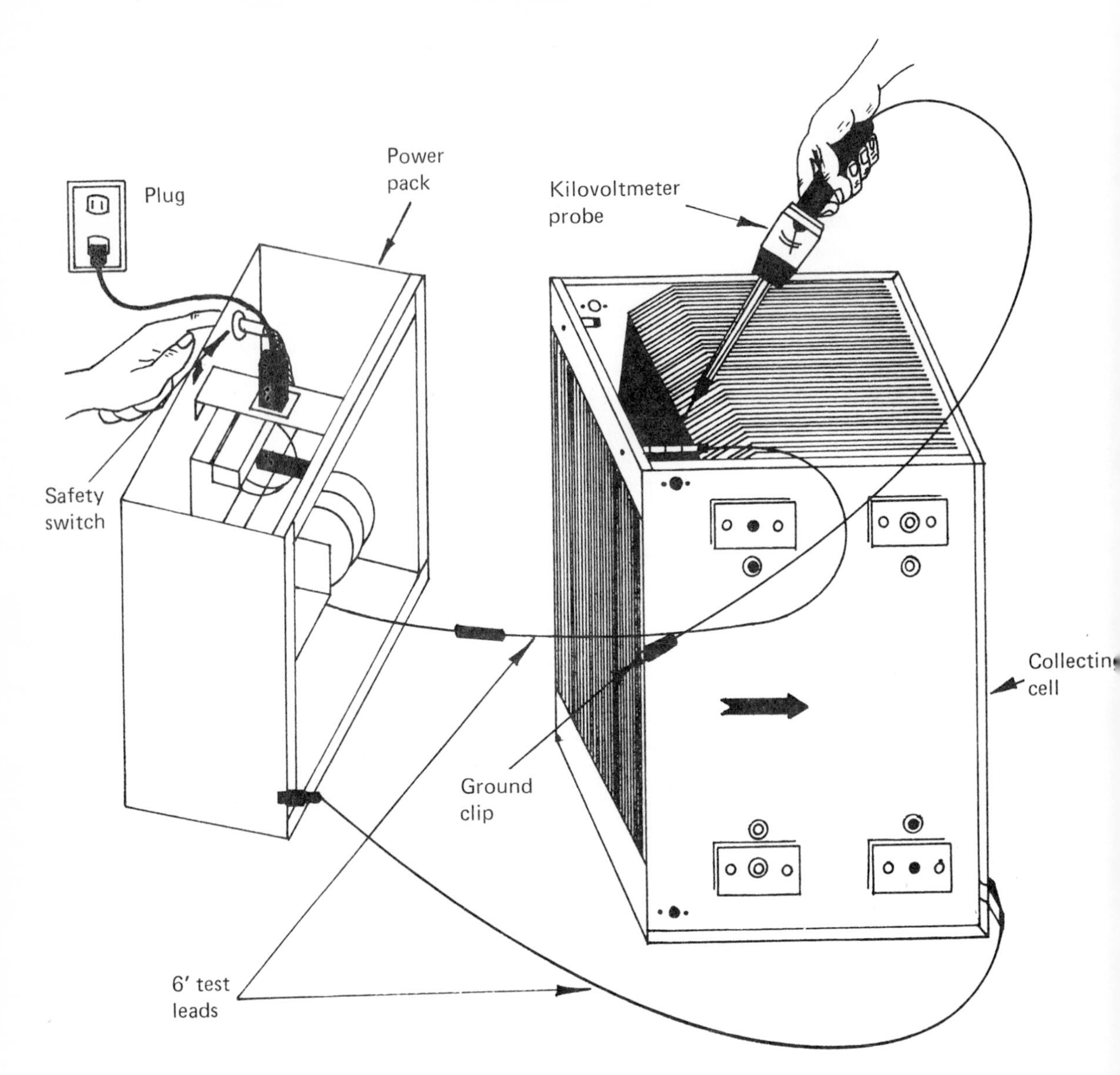
Power
pack
Plug
Kilovoltmeter
probe
Safety
switch
Collectin
cell
Ground
clip
6' test
leads

Figure 9.17.

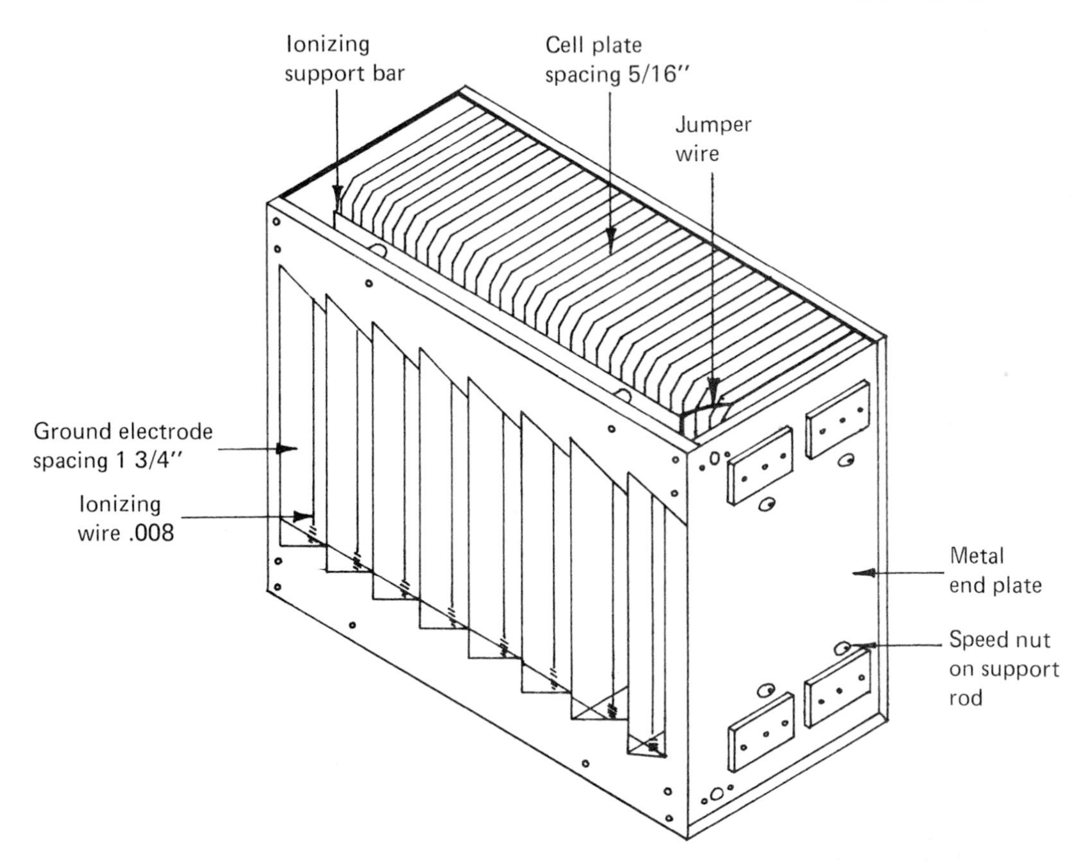

Figure 9.18. Super compact ionizing collecting cell.

Replace the screw and insulator that is being checked and proceed to another insulator where the same testing procedure should be applied. When the screw is removed from a defective insulator the reading on the kilovoltmeter-probe will return to normal. This insulator should be replaced.

8. With the ohmmeter scale set at Rx 100,000, a check for continuity can be made between the ionizing bar and the ground electrodes of the ionizing section. The reading should be infinite ohms. Any reading but infinite ohms proves that a foreign object is in the ionizing section or that an ionizing wire has broken and is touching a ground electrode or there is a defective insulator. The ohmmeter test is not totally conclusive. Occasionally, a very fine crack in an insulator will result in the insulator breaking down under a high voltage condition. In this particular case, the ohmmeter test for shorting would be inadequate.

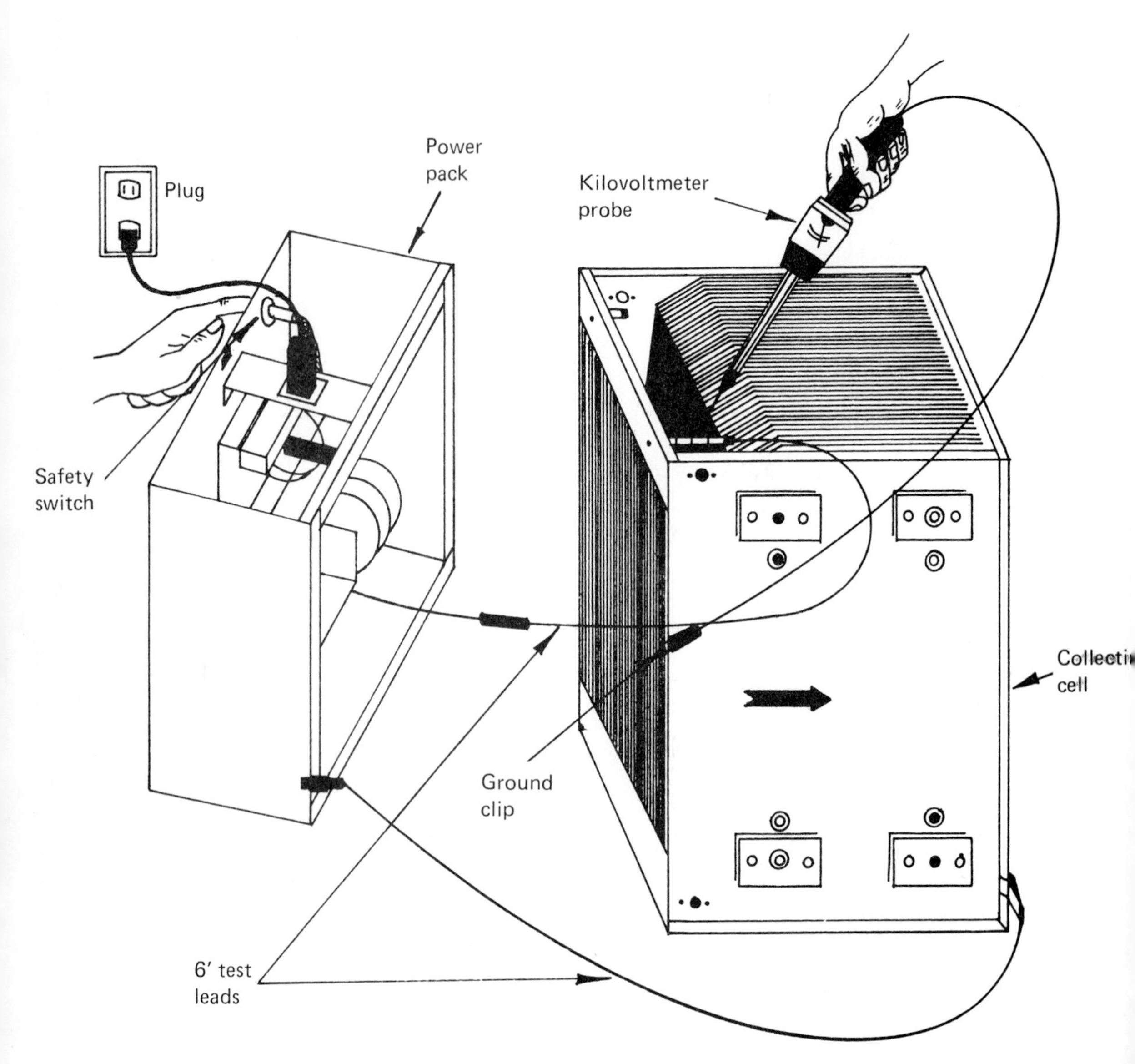
Power
pack
Plug
Kilovoltmeter
probe
Safety
switch
Collecti
cell
Ground
clip
6′ test
leads

Figure 9.19.

Servicing the Power Pack

If the service problem has not been located in the cell, the next area to investigate is the power pack.

1. Establish the Service Ready Stage. Connect Power supply cord, voltage selector switch turned to position #3 and safety switch ready to be depressed to temporarily energize the air cleaner.
2. Using the kilovoltmeter-probe depress the safety switch and read the voltage. If the voltage is below 9,500 volts, and the A.C. ammeter is in service, there is a faulty high voltage component.

There are three major high voltage components in the power pack. The easiest way to determine if each of these components is functioning correctly is to isolate each one and to check it separately.

A. **Transformer Check**
Remove the selenium rectifiers. Depress the safety switch to energize the power pack and observe the A.C. ammeter. If it returns to the *white on area,* the transformer is functioning properly. If it returns to a position that is very low in the *white on area,* the transformer is defective and must be replaced.

B. **Rectifier Test**
Rectifiers can be field tested by inserting a pencil in the selenium end of the rectifiers (arrow points toward this end). If the rectifier is good, the selenium will slide easily within ths tube against the spring in the opposite end. If the rectifier is defective, the selenium will expand against the side of the tube and cannot be moved.

C. **Capacitor Check**
If the checking procedures listed before have still not located the source of trouble, replace the capacitor assembly. Occasionally the power pack ammeter will indicate that the unit is operating correctly, when in fact, it is not. In this situation the indicating needle is registering in any portion of the *white on area.* To determine if the unit is functioning with the correct output voltage, it is necessary to check the voltage at the thru-bushing with the kilovoltmeter-probe in the following manner:
   1. Attach the black clip from the kilovoltmeter-probe's lead to the power pack housing. The metal end of the probe should touch the thru-bushing screw. Readings on the kilovoltmeter-probe should be between 7,500 and 8,400 volts.
   2. If the voltages are lower than 7,500, proceed as follows:
      a. Check the rectifiers to see that they match the directional arrows on the high voltage board.
      b. Capacitor open—replace the capacitor assembly.

The first indication that a super compact unit is not working correctly is when the A.C. ammeter is in the *black off area*. This indicates that 120 volt power is not being supplied to the transformer. To determine that voltage is or is not available, proceed in the following manner:

1. Check the supply voltage with a test light or with the volt-ohmmeter set at the 300v AC scale. It should read 110-120v A.C. if the light continues to glow, indicating an adequate power supply. Proceed to individaully check the primary components.
2. Check each of the primary components shown with the test light. The test light will glow until you pass the defective component.

An alternate method of checking the primary components in the power pack to discover a defective component is to use a volt-ohmmeter. Before attempting this method, remove the 110 volt power cord. The components to be checked with a volt-ohmmeter are as follows:

1. Remove the fuse and check for continuity with the volt-ohmmeter on the Rx1 ohms scale. The reading should be zero ohms. If the reading is zero, the fuse is good and should be removed for the rest of the tests. If the fuse if bad, replace it.
2. *Voltage Selector Switch*
   Connect the volt-ohmmeter across points one and three. With the meter set on Rx 10, rotate the knob and read the meter. The readings should be as follows:

Off	**Infinite Ohms**
1	75 **Ohms**
2	40 **Ohms**
3	o **Ohms**

If the readings are not as shown, replace the switch and resistors.

3. *Safety Switch*
   Connect the volt-ohmmeter across the switch and depress the switch several times. The meter should deflect to zero ohms (on the Rx1 scale) when the switch is depressed and to infinite ohms (on the Rx 100,000 scale) when it is released.
4. ***A.C. Ammeter—Open***
   With the volt-ohmmeter on the Rx 1 scale, connect the leads of the meter across the terminals of the A.C. ammeter. A reading of approximately 2.5 ohms indicates that the meter is operating properly. A reading of infinite ohms shows the meter's winding is open and the meter should be replaced. An alternate check of the ammeter can be performed by using the kilovoltmeter-probe. Reconnect the 110 volt power cord and the fuse. Short out the meter (connect a wire from one terminal to the other) and depress the safety switch to energize the unit. Use the kilovoltmeter-probe to determine if voltage has been restored to the output of the power pack. If the voltage reads correctly (7,500-8,400) the A.C. ammeter is defective and should be replaced.

Determining that the collecting cell is functioning after correcting a power pack service problem. There is a possiblity that two separate service problems may have occurred in the electronic air cleaner: one located in the power pack and the other in the collecting cell. If the problem has been located in the power pack and has been satisfactorily corrected, the collecting cell should be checked to determine that there is no problem there. This is done in the following manner.

1. Attach the high voltage lead to the thru-bushing.
2. Depress the safety switch to temporarily energize the air cleaner
3. Observe the A.C. ammeter. If the indicating needle is in the *black service area* to the right of the *white on area.* The collecting cell contains a service problem.
4. Proceed to check out the collecting cell.

Determining if the A.C. ammeter is defective. There is a slight possibility that the ammeter would indicate that the electronic air cleaner requires service when, in fact, the ammeter itself is defective this could be caused by poor workmanship in the manufacture of the ammeter or user tampering.

The ammeter in this condition would show the indicating needle in the service area. To check out the ammeter, the following procedure should be used:

1. Establish the service ready stage. (Connect power supply cord, voltage selector switch turned to position #3 and safety switch ready to be depressed to temporarily energize the air cleaner).
2. Using the kilovoltmeter-probe check the voltage to the cell. Attach black clip lead from the kilovoltmeter-probe to the power pack housing and touch the metal end of the probe to the thru-bushing.
3. Depress the safety switch to temporarily energize the air cleaner.
4. Observe the voltage shown on the kilovoltmeter-probe. If the voltage reading is normal (7,500-8,400 Volts) the problem is a defective ammeter which should be replaced with a new ammeter.

Additional Service and Maintenance Tips

1. Initial installation.
   Upon initial energization of the unit, a low voltage and high current condition may exist. This is due to the high ionizing current draw of the new cell. This is a normal condition and will correct itself within approximately 48 hours of operation.
2. Continuous arcing problem
   a. High voltage-unit should be operated at the highest voltage for maximum efficiency. This voltage is achieved when the voltage selector switch is in position #3. If arching occurs, rotate voltage selector switch to position #2. If arcing still occurs use position #1.
   b. Bent or loose plates—can be corrected by straightening plates and reconnecting plates to bars used to hold plates in place.
   c. Check hand valve—should be checked to insure that there is no water leaking into cell.
   d. Maintenance of collecting cell—a cell should be removed from its cabinet at least once a year and manually washed with hot water and a dish washing detergent or DAX-W. After washing with a detergent, rinse with hot water and drain. Replace the cell and allow the furnace blower to operate approximately one hour before re-energizing the power pack. DAX-W is available from distributors.

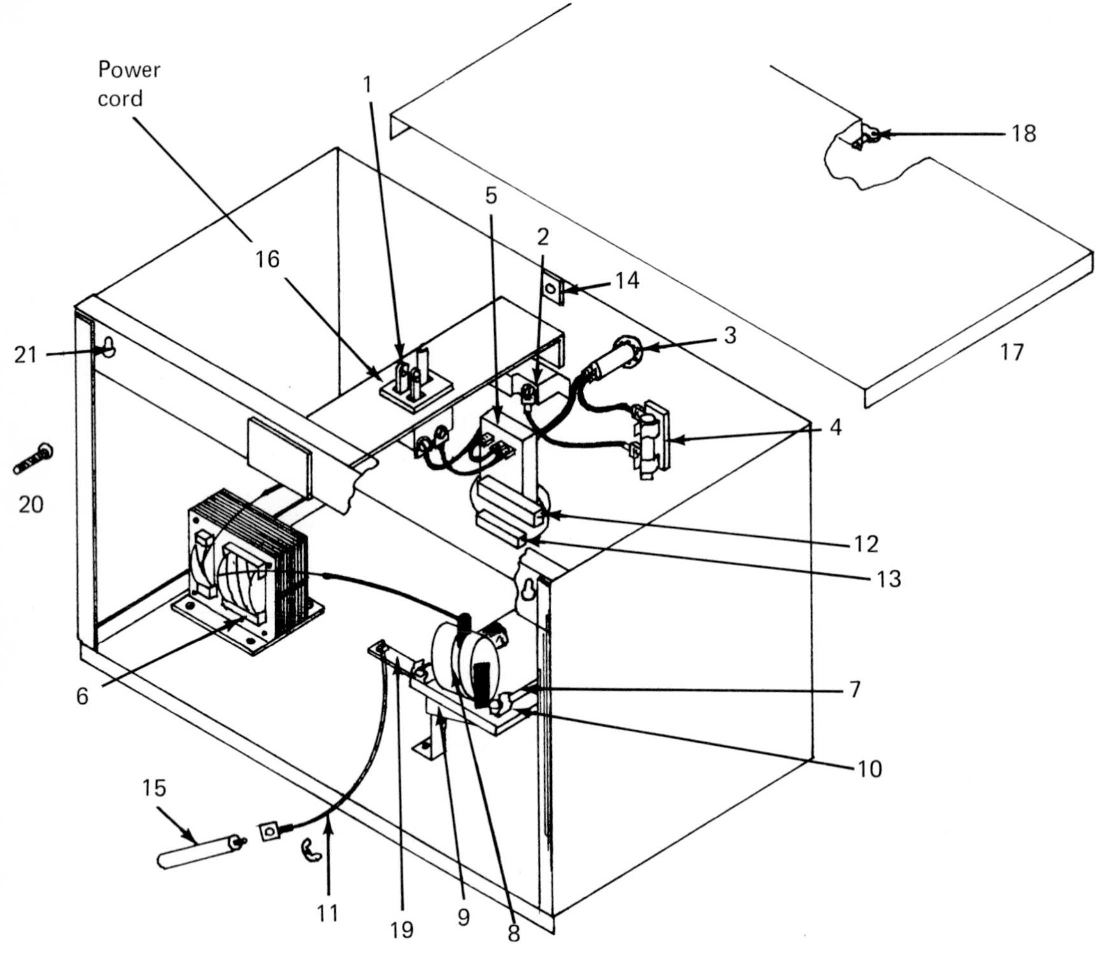

1. Male receptacle
2. A.C. ammeter
3. Safety switch
4. Fuse
5. On/off switch
6. Transformer
7. Selenium rectifiers
8. Capacitors
9. Bleed off resistor
10. High voltage board
11. High voltage lead
12. 75 OHM resistor
13. 40 OHM resistor
14. Lid catch
15. Thru bushing
16. Power supply cord
17. Power pack lid
18. Quarter turn fastener
19. High voltage bracket
20. 1/4 20 Mounting screw
21. Key-hole slot

Figure 9.20. Super compact power pack general arrangement.

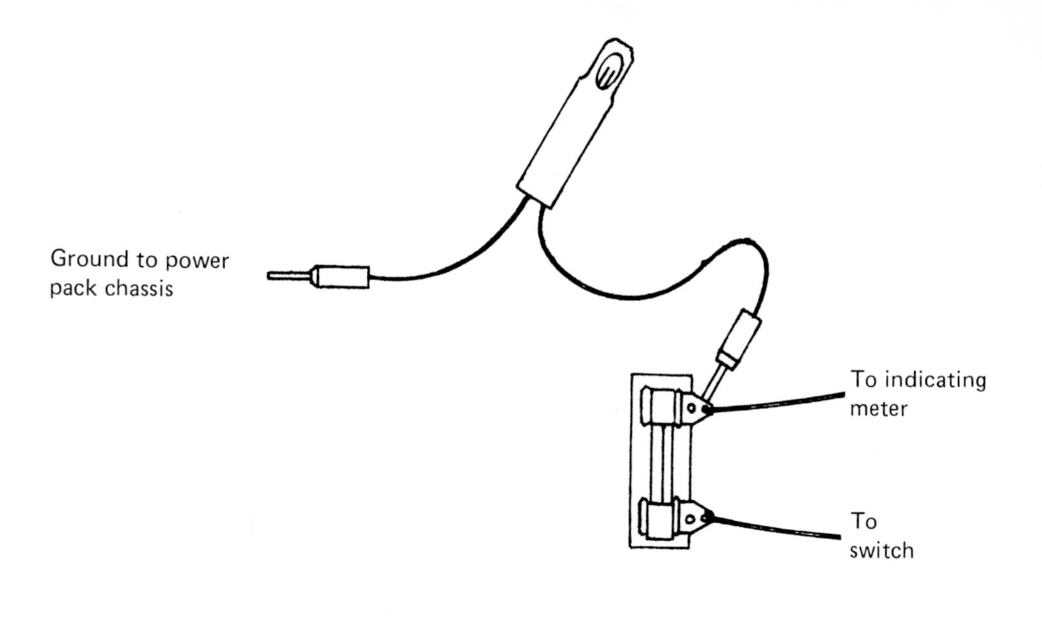
Ground to power
pack chassis
To indicating
meter
To
switch

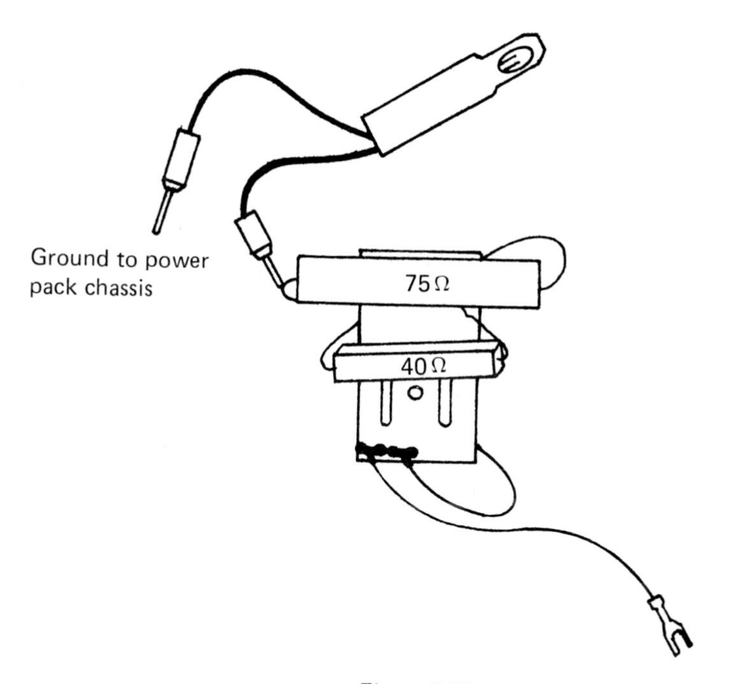
Ground to power
pack chassis
75Ω
40Ω

Figure 9.21.

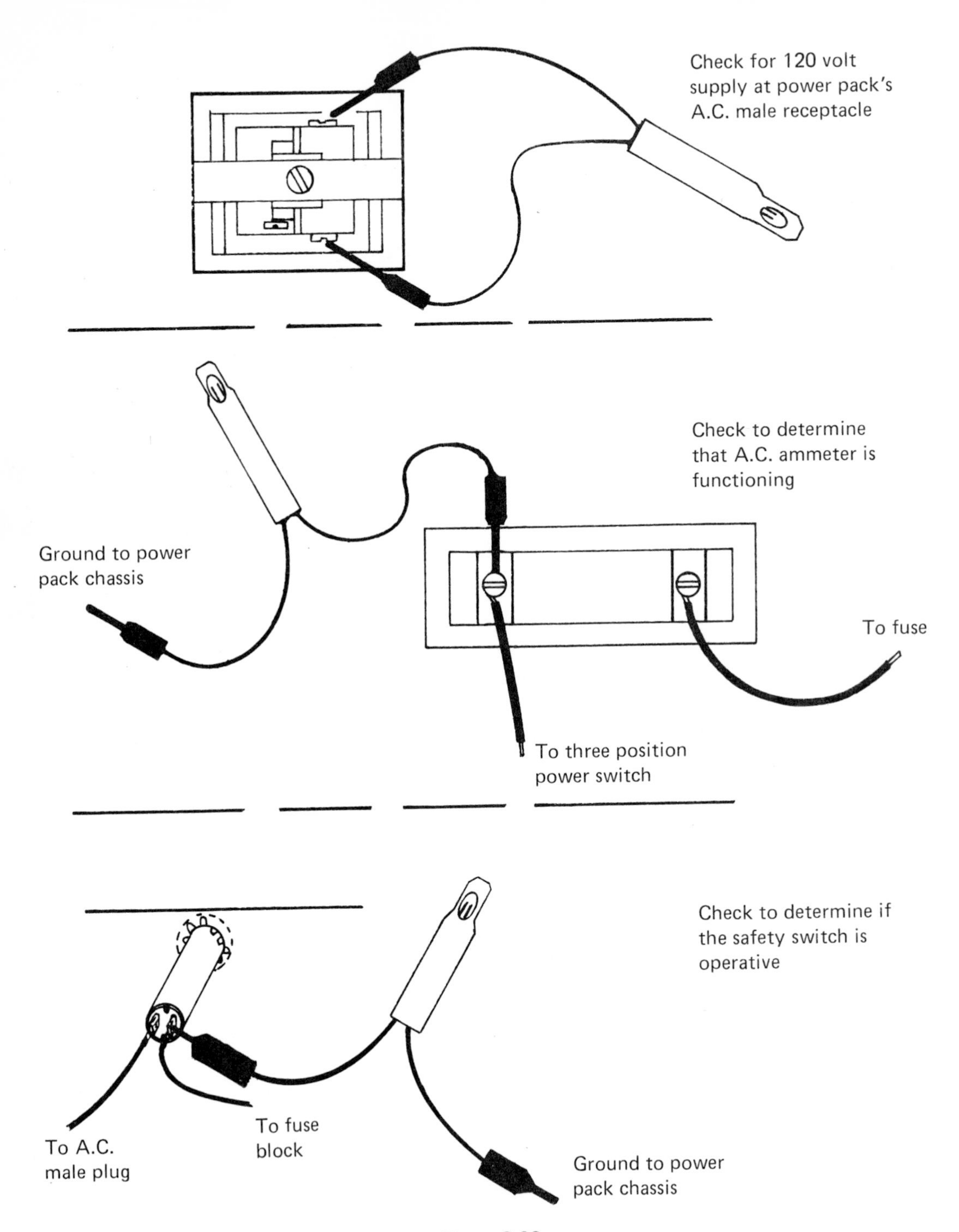
Check for 120 volt
supply at power pack's
A.C. male receptacle
Check to determine
that A.C. ammeter is
functioning
Ground to power
pack chassis
To fuse
To three position
power switch
Check to determine if
the safety switch is
operative
To A.C.
male plug
To fuse
block
Ground to power
pack chassis

Figure 9.22.

Chapter 10
HUMIDIFIERS

Humidity affects your health, comfort, and property. In the winter, the heated home without humidification is about 10 to 15% relative humidity—dangerous to your health according to health authorities. The Sahara and Gobi Deserts—two of the world's driest natural areas have a relative humidity of about 25%. The ideal relative humidity level should be 30% to 40%, contingent somewhat on outdoor and indoor temperatures. So in the winter, the house that is without a humidifier can be drier than a desert.

There's absolute humidity and relative humidity. Absolute humidity is the term used to denote the amount of water vapor in a given amount of air by weight or measurement in grains of water per cubic foot of air. Relative humidity is the amount of water vapor in the air in relation to the amount the air could contain at the same temperature expressed in percentages. When air is completely saturated, or holding as much water vapor as possible, the relative humidity is 100%. When air contains half as much water vapor as it possibly can, the relative humidity is 50%.

In a house there is a complete change of air in a winter-heated home about every one to two hours. This changing air occurs through the opening of doors and leakage around windows and joints. This leakage is called infiltration. This infiltration is caused by your furnace when your furnace is burning. It takes 10 to 15 cubic feet of air to burn 1,000 B.T.U's. This air must be replaced; if not, your furnace and you will lack oxygen, and your furnace will not want to burn right.

So even if it were possible to "seal off" against humidity loss, it would be undesirable. Fresh air is needed constantly; moisture is also needed constantly in the winter. And the vast amounts of outside cold air coming in and being heated reduces the inside moisture considerably. Some will open a window at night to relieve distress caused by dry heated air; in fact, the opening of a window will cause the dry air condition in the home worsen. For example, outside air at zero degrees and 75% humidity is reduced to under 5% humidity when heated to 70 degrees—very low.

Some humidity is regained through water vapors given off in cooking, laundering, bathing, and washing. But all this water vapor added together, at best, is only a few pints a day, which is inadequate. Since warm air is capable of holding more water vapor than cool air when your furnace goes on, the cool air is heated and its capacity for holding water vapor goes way up; therefore, the humidity goes way down. When air is of low humidity, it takes moisture from everything it contacts—for example, your skin, respiratory passages, hair, and your furnishings. The result—a dried house and everything in it.

The average house requires about seven gallons of water evaporation during each day of winter heating for proper humidification. These are some of the things that proper humidity can do for the home and its occupants. Because unhumidified-heated air takes moisture from your skin and body, a higher temperature is usually required for personal comfort resulting in excessive fuel consumption and higher fuel bills. In essence you will be more comfortable at 65 °F with proper humidity, than 70 °F with little or no humidity.

Proper humidity will also help walls and plaster from cracking—electrical shocks from static electricity—woodwork, cabinets, and drawers from shrinking and cracking—paint and wallpaper from chipping, flaking, or peeling. Proper humidity will also help rugs and all fabrics to wear longer—to keep glued joints and veneers of furniture from separating—flooring from creaking—linoleums and floor tile from coming loose—leather and books from aging prematurely—paintings from stiffening—house plants from wilting or sometimes dying.

Determining Humidity Requirements. Output required is determined by the cubic feet of heated air space to be humidified and the estimated amount of air changes per hour based on construction. After this has been determined, select the factor from Chart #1 which will determine the realtive humidity desired at 70 °F room temperature. Next, multiply the cubic feet by the amount of air changes, then by the factor. Divide the answer by 7,000 to convert from grains of moisture to pounds of water per hour.

Example. An 18,000 cubic foot house has one air change per hour. How much water must be added per hour to maintain 35% relative humidity?

Formula. Cubic foot X air change X factor divided by 7,000 equals desired output.

Answer. $\frac{18,000 \times 1 \times 2}{7,000}$ = 5.15 lb. per hr. or 15 gals. per day.

Chart 10.1.

Relative Humidity Desired %	Factor
20	0.83
25	1.13
30	1.65
35	2.00
40	2.44
45	2.81

Chart 10.2.

Any Change in Temperature Changes the
Right Hand Column Readings

Temperature 70°F	
Relative Humidity	**Absolute Humidity**
Maximum One cu. ft. of Air Can Hold 70°F	Grains of Water per cu. ft. of Air
100%	8.0
90%	7.2
80%	6.4
70%	5.6
60%	4.8
50%	4.0
40%	3.2
30%	2.4
20%	1.6
10%	.8

Chart 10.3.

Indoor Relative Humidity Which Occurs as Cold Outdoor Air Is Heated to 75°F Without Adding Moisture

Outdoor Temp. Degrees F	Outdoor Relative Humidity Percent	Relative Humidity in Home Without Humidification Percent
−10	100 70	3 2
0	100 70	4 3
10	100 70	7 4
20	100 70	11 8
30	70 50	13 9

Chart 10.4.

Size of House Cu. Ft.	Outside Temp. Degrees F	Recommended Relative Humidity Setting %	Moisture Average House Gal-Day	Required Tight House Gal-Day
8.000	0	25	2.9	1.4
	10	30	3.4	1.8
	20	35	3.7	1.9
	30	35	3.2	1.6
	40	40	3.6	1.8
12.000	0	25	4.3	2.1
	10	30	5.2	2.5
	20	35	5.4	2.6
	30	35	5.0	2.4
	40	40	5.3	2.5
16.000	0	25	5.8	2.8
	10	30	6.8	3.4
	20	35	7.5	3.7
	30	35	6.4	3.3
	40	40	7.4	3.6
20.000	0	25	7.2	4.3
	10	30	8.4	5.1
	20	35	9.1	5.6
	30	35	8.1	4.8
	40	40	9.0	5.5
24.000	0	25	8.7	4.3
	10	30	10.2	5.0
	20	35	11.2	5.6
	30	35	9.7	4.8
	40	40	11.1	5.5

*Chart based on information supplied by technical services department national warm air heating and air conditioning association.

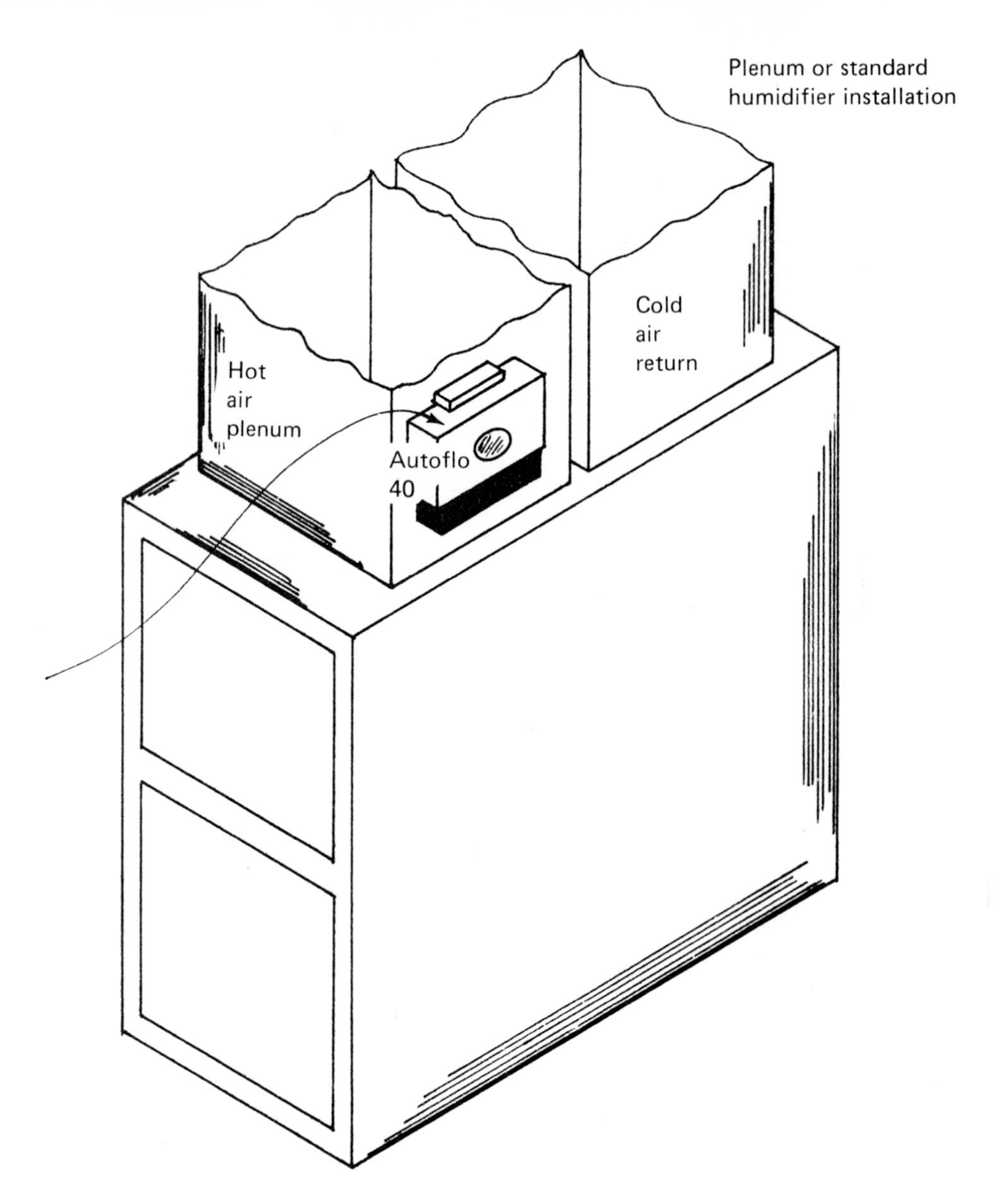

Figure 10.1. Typical installation of humidifier on a low-boy furnace.

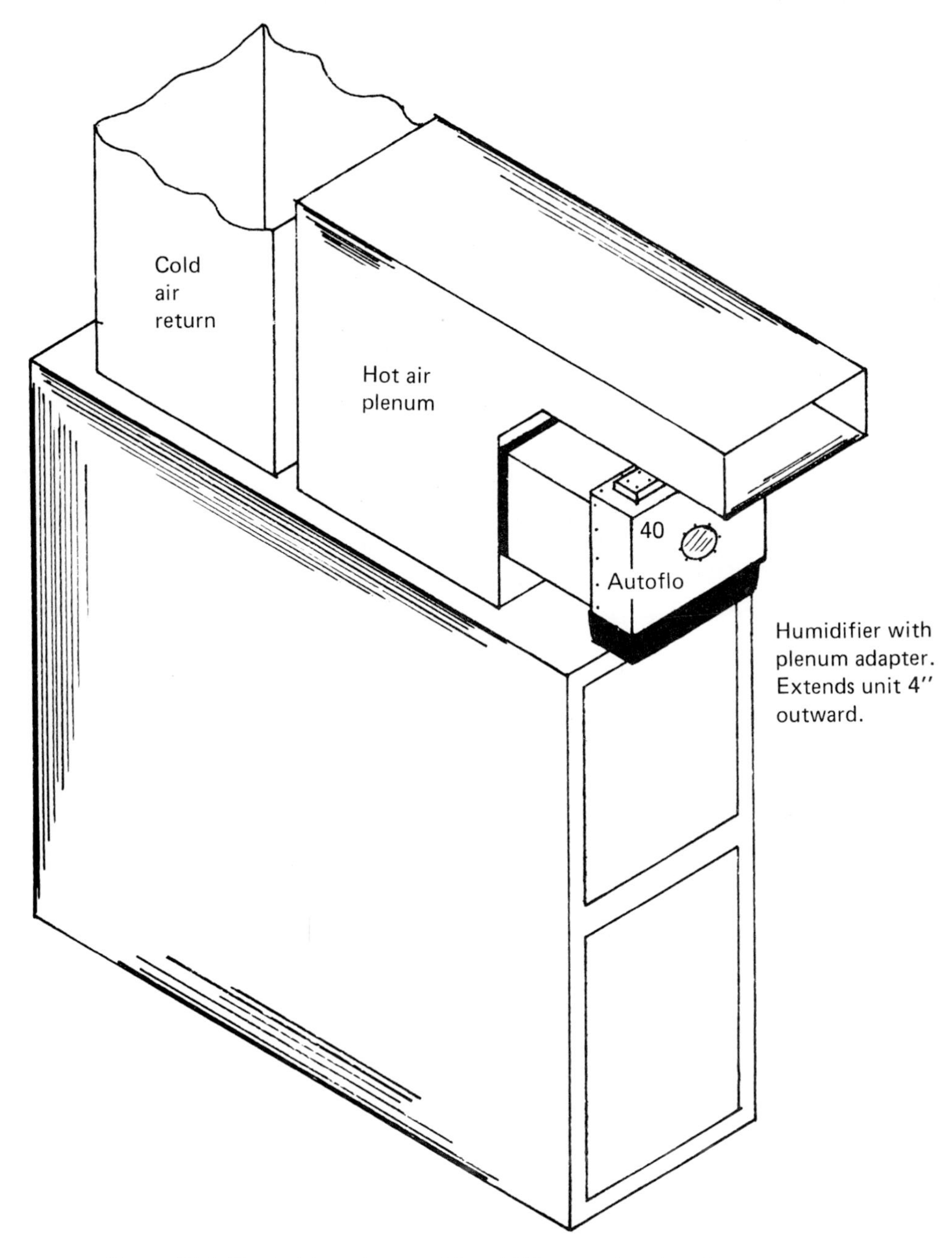

Figure 10.2. Typical installation of humidifier on a low-boy furnace.

For cluttered furnace unit. May be installed away from furnace in the hot air duct with an adapter.

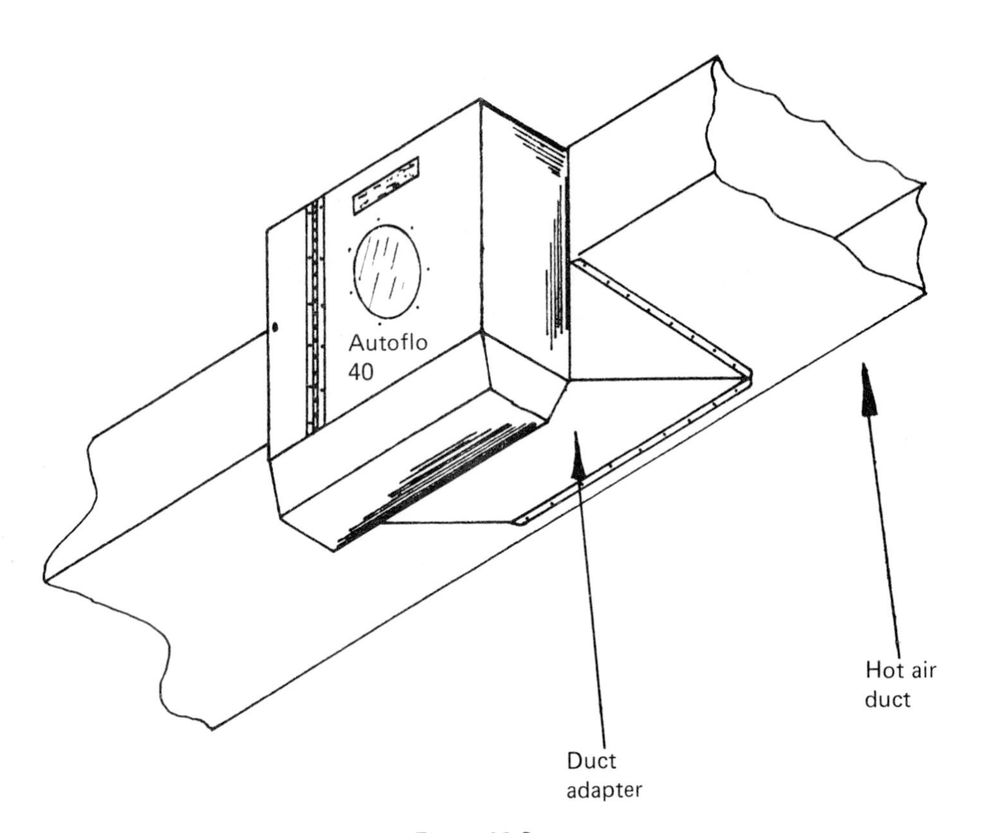

Figure 10.3.

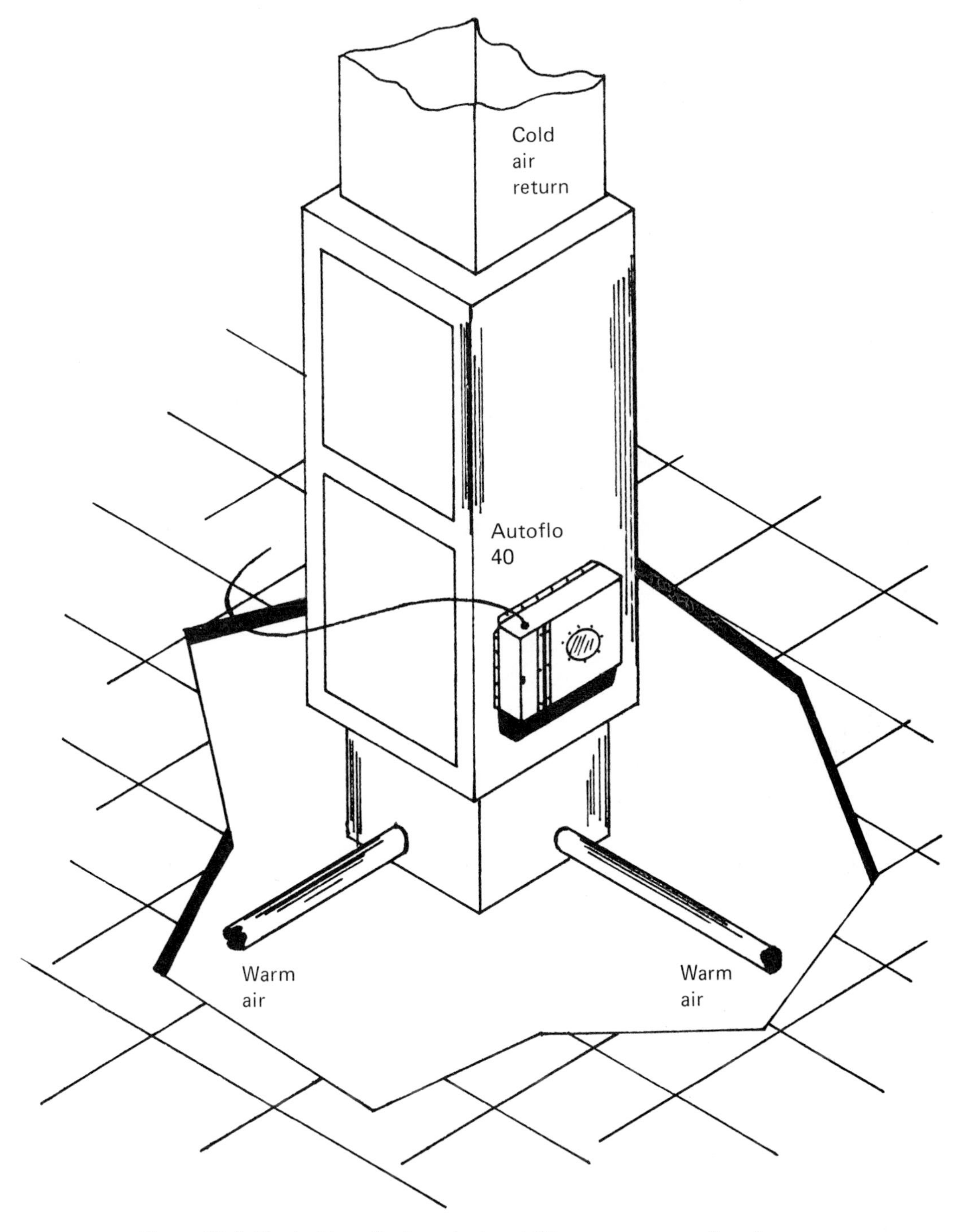

Figure 10.4. Typical installation of a humidifier on a counter-flow furnace.

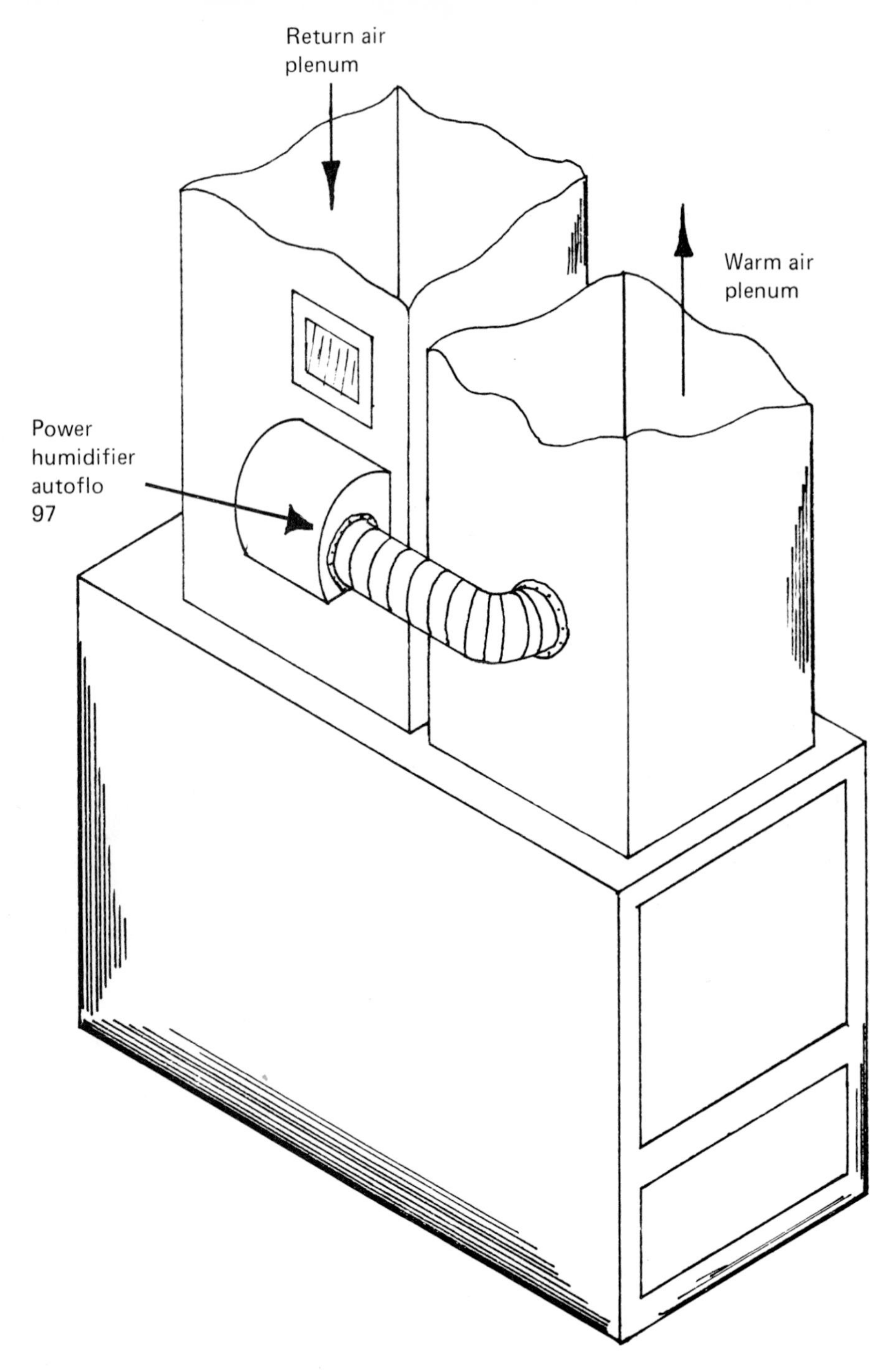

Figure 10.5. Typical installation of a humidifier on a low-boy furnace.

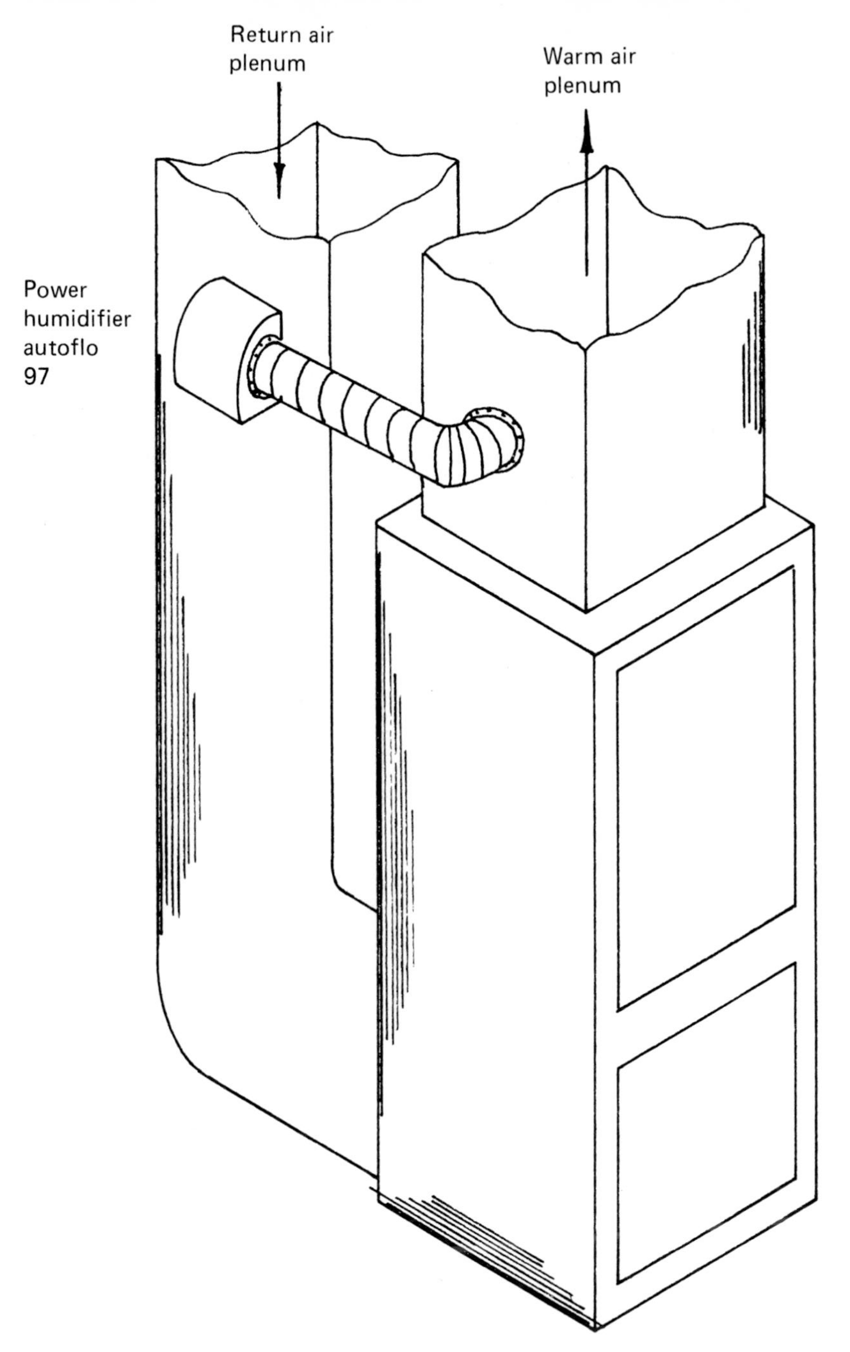

Figure 10.6. Typical installation of a humidifier on a high-boy furnace.

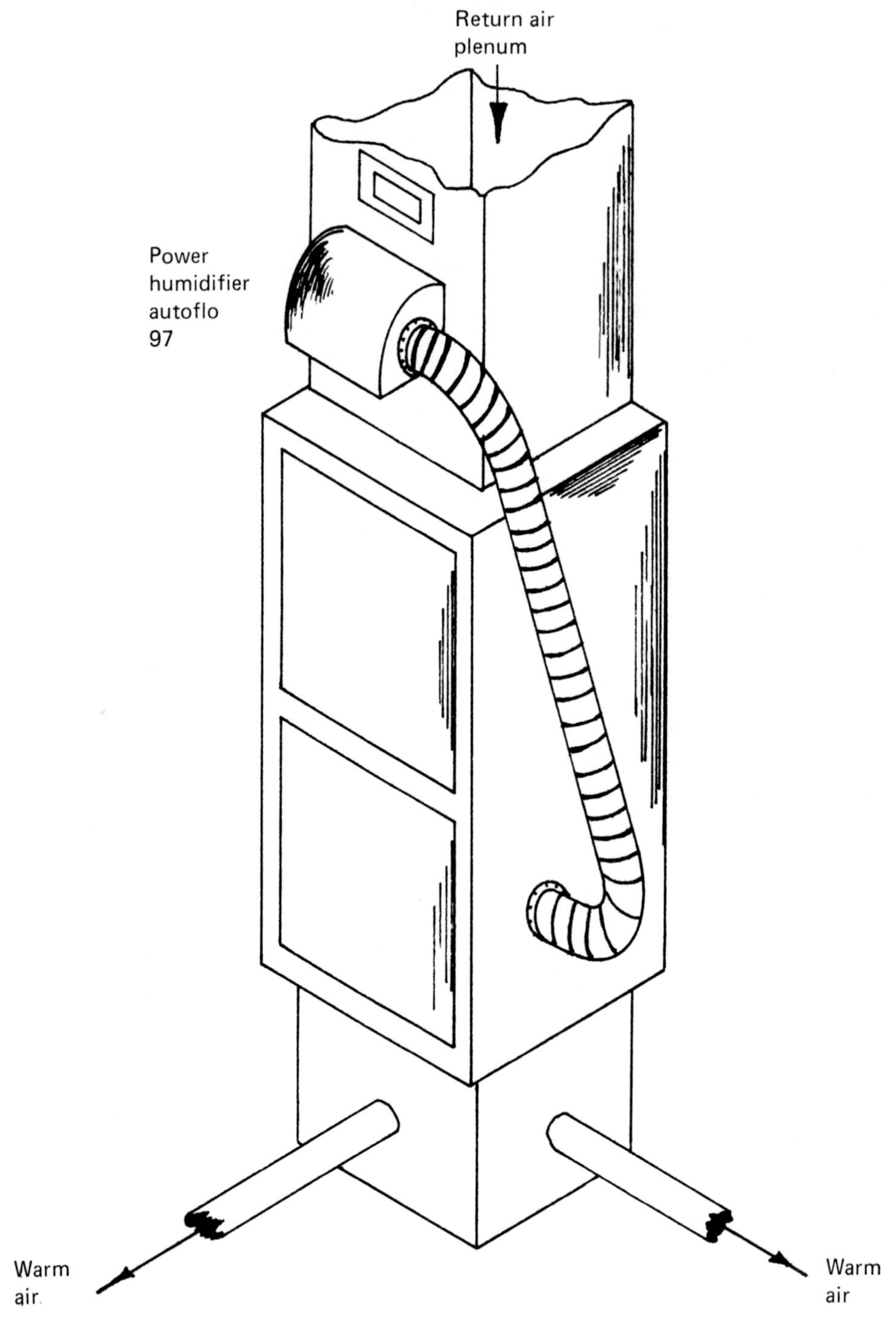

Figure 10.7. Typical installation of a humidifier on a counter-flow furnace.

APPENDIX

GENERAL INFORMATION

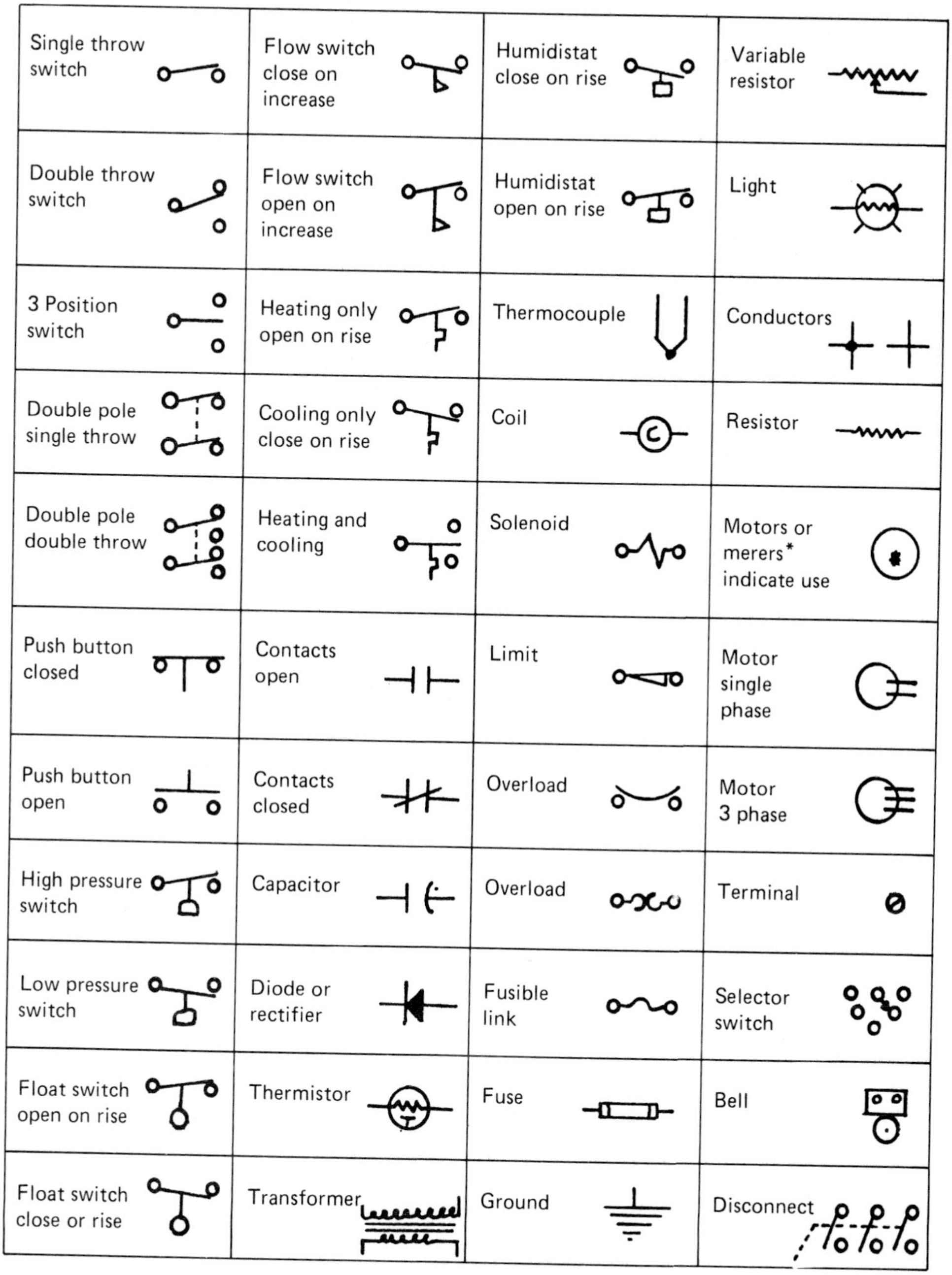

Single throw switch	Flow switch close on increase	Humidistat close on rise	Variable resistor
Double throw switch	Flow switch open on increase	Humidistat open on rise	Light
3 Position switch	Heating only open on rise	Thermocouple	Conductors
Double pole single throw	Cooling only close on rise	Coil	Resistor
Double pole double throw	Heating and cooling	Solenoid	Motors or merers* indicate use
Push button closed	Contacts open	Limit	Motor single phase
Push button open	Contacts closed	Overload	Motor 3 phase
High pressure switch	Capacitor	Overload	Terminal
Low pressure switch	Diode or rectifier	Fusible link	Selector switch
Float switch open on rise	Thermistor	Fuse	Bell
Float switch close or rise	Transformer	Ground	Disconnect

Figure A1.

Horsepower—Ampere Table

Approximate Horsepower	120 Volts A.C.		240 Volts A.C.	
	Full Load	Locked Rotor	Full Load	Locked Rotor
1/10	3.0	18.0	1.5	9.0
1/8	3.8	22.8	1.9	11.4
1/6	4.4	26.4	2.2	13.2
1/4	5.8	34.8	2.9	17.4
1/3	7.2	43.2	3.6	21.6
1/2	9.8	58.8	4.9	29.4
3/4	13.8	82.8	6.9	41.4
1	16.0	96.0	8.0	48.0
1 1/2	20.0	120.0	10.0	60.0
2	24.0	144.0	12.0	72.0
3	34.0	204.0	17.0	102.0

Figure A2.

Sheet-Metal Thickness

Gauge No.	Uncoated Low-Alloy Steel*	Aluminum Brass and Copper
28	.015-1/64"	.012
26	.018	.016-1/64"
24	.024	.020
22	.030	.025
20	.036-1/32"	.032-1/32"
18	.048-3/64"	.040
16	.060-1/16"	.051
14	.075-5/64"	.064-1/16"
12	.105-7/64"	.081-5/64"

*Galvanized and other coated steel slightly thicker.

Figure A3.

Wire Ampere Ratings

Size AWG MCM	Maximum Current Rating, Amperes*			
	Type R, RW, RV, T, TW	Type RH, RUH, THW	Type A1, A1A	Type A AA
Solid Conductor				
No. 14	15	15	30	30
12	20	20	40	40
10	30	30	50	55
8	40	45	65	70
Stranded Conductor				
6	55	65	85	95
4	70	85	–	–
3	80	100	–	–
2	95	115	–	–
1	110	130	–	–
0	125	150	–	–
00	145	175	–	–
000	165	200	–	–
0000	195	230	–	–
250 MCM	215	255	–	–
300	240	285	–	–
350	260	310	–	–
400	280	335	–	–
500	320	380	–	–

Figure A4.

Power and Heat

1 BTU . . . 776 ft.-lb.–0.293 watt–hr.–262 cal.
1 CAL . . . 0.003968 BTU 0.0011619 watt.–hr.
1 BTUH . . . 0.293 watts.–4.2 cal/min
12.000 BTU . . . 1-ton refrigeration
1 watt . . . 3.413 BTUH
1 watt.-hr . . . 3.413 BTU
1 kw. (1,000 watts) 3413 BTUH
1 kw-hr . . . 3413 BTU
1 HP . . . 0.746 kw.-2647 BTUH.-33.000 ft.-lb/min
1 BOHP . . . 33,475 BTUH.

Figure A5.

POWER FACTOR

The power-factor of any a-c circuit is equal to the true power in watts divided by the apparent power in volt-amperes which is equal to the cosine of the phase angle, and is expressed by

$$\text{p.f.} = \frac{EI \cos \theta}{EI} = \cos \theta$$

Where

p.f. = the circuit load power factor,

$EI \cos \theta$ = the true power in watts,

EI = the apparent power in volt-amperes,

E = the applied potential in volts,

I = load current in amperes.

Therefore

in a purely resistive circuit,

$\theta = 0^\circ$ and p.f. = 1

and in a reactive circuit,

$\theta = 90^\circ$ and p.f. = 0

and in a resonant circuit

$\theta = 0^\circ$ and p.f. = 1

Ohms Law for D-C Circuits

The fundamental Ohms law formulas for d-c circuits are given by,

$$E = IR, \qquad I = \frac{E}{R}, \qquad R = \frac{E}{I}, \qquad P = EI.$$

Where

I = current in ampers,

R = resistance in ohms,

E = potential across R in volts,

P = power, in watts.

Ohms Law for A-C Circuits

The fundamental Ohms law formulas for a-c circuits are given by

$$I = \frac{E}{Z}, \qquad Z = \frac{E}{I}, \qquad E = IZ, \qquad P = EI \cos\theta$$

Where

- I = current in amperes,
- Z = impedance in Ohms,
- E = volts across Z,
- P = power in watts,
- 0 = phase angle in degrees

Ohm's Law Formulas for A.C. Circuits

Known Values	Formulas for Determining Unknown Values			
	I	Z	E	P
I and Z			IZ	$I^2 z \cos\theta$
I and E		E/I		IE $\cos\theta$
I and P		$\frac{P}{I^2 \cos\theta}$	$\frac{P}{I \cos\theta}$	
Z and E	E/Z			$\frac{E^2 \cos\theta}{Z}$
Z and P	$\sqrt{\frac{P}{z \cos\theta}}$		$\sqrt{\frac{pz}{\cos\theta}}$	
E and P	$\frac{p}{E \cos\theta}$	$\frac{E^2 \cos\theta}{P}$		

Ohm's Law Formulas for D.C. Circuits

Known Values	Formulas for Determining Unknown Values			
	I	R	E	P
I and R			IR	I^2R
I and E		E/I		EI
I and P		P/I^2	P/I	
R and E	E/R			E^2/R
R and P	$\sqrt{\frac{P}{R}}$		$\sqrt{PR}$	
E and P	P/E	E^2/P		

PHASE ANGLE

The phase angle is defined as the difference in degrees by which current leads voltage in a capacitive circuit, or lags voltage in an inductive circuit, and in series circuits is equal to the angle whose tangent is given by the

$$\text{ratio } \frac{X}{R} \text{ and is expressed by arc tan } \frac{X}{R}$$

Where

X = the inductive or capacitive reactance in ohms,

R = the non-reactive resistance in ohms,

of the combined resistive and reactive components of the circuit under consideration.

Therefore,

in a purely resistive circuit, $\theta = 0°$

in a purely reactive circuit, $\theta = 90°$

and in a resonant circuit, $\theta = 0°$

also when

$\theta = 0°$, $\cos\theta = 1$ and $P = EI$, $\theta = 90°$, $\cos\theta = 0$ and $P = 0$.

Degrees 0.0175 = radians.
1 radian = 57.3°.

Index